TWIN TRACKS

THE UNEXPECTED ORIGINS OF THE MODERN WORLD

James Burke

SIMON & SCHUSTER
New York London Toronto Sydney Singapore

SIMON & SCHUSTER
Rockefeller Center
1230 Avenue of the Americas
New York, NY 10020

For information regarding special discounts for bulk purchases,
please contact Simon & Schuster Special Sales:
1-800-456-6798 or business@simonandschuster.com

Designed by Jaime Putorti

Manufactured in the United States of America

10 9 8 7 6 5 4 3 2 1

Library of Congress Cataloging-in-Publication Data

Burke, James, 1936–
Twin Tracks : the unexpected origins of the modern world / James Burke.
p. cm.
Includes bibliographical references and index.
1. Civilization, Modern—History. I. Title.

CB358.B88 2003
909.08—dc21

2003052628

ISBN 0-7432-2619-4

ACKNOWLEDGMENTS

I should like to express my grateful thanks to Carolyn Doree for her peerless assistance in research.

To Madeline

CONTENTS

Introduction 1
How to Read This Book 3

1. 1804: Attack on Tripoli to Fish Sticks 5
2. 1760: Fake Epic to Organ Transplants 15
3. 1805: Battle of Trafalgar to Laser 25
4. 1726: Encyclopedia to Vitamins 35
5. 1792: Juniper Hall to Jet Aircraft 45
6. 1750: Smallpox to Big Bang 55
7. 1784: Sanskrit to Cybernetics 65
8. 1610: *Santa Catharina* to Spectroscopy 75
9. 1686: Political Jingle to Nylon 85
10. 1703: Kit-Kat Club to Sunglasses 95
11. 1770: Falklands War to Television 105
12. 1724: Stone Age Boy to Photocopier 115
13. 1745: Leyden Jar to Clingwrap 125
14. 1790: *Philadelphia General Advertiser* to Chemotherapy 135
15. 1664: Lens Grinder to Hairdressing 145
16. 1773: Boston Tea Party to Contact Lenses 155
17. 1742: Bow Street, London, to Bar Code 165
18. 1739: The Grand Tour to Liquid Crystal Display 175
19. 1795: Man in the Iron Mask to Hovercraft 185
20. 1673: Siege of Maastricht to Vending Machines 195
21. 1786: *The Marriage of Figaro* to Stealth Fighter 205
22. 1780: Edinburgh Oyster Club to DNA 215
23. 1770: Church Sermon to Helicopter 225
24. 1771: Pottery to Neon Signs 235
25. 1676: Theology to Skyscraper 245

Bibliography 255
Index 262

INTRODUCTION

Heading confidently for Japan, Columbus unexpectedly bumped into America, and Western knowledge went down the tubes. What was the New World doing there, when it didn't figure in the Bible or Aristotle? Not to mention its thousands of never-before-seen plants and animals. Intellectual panic followed. If the classical authorities were wrong about something as big as this, whose word could you trust any more? A hundred years of global exploration later, the problem had grown too big to ignore, so in 1619 René Descartes came up with a way to verify data through methodical doubt and reductionism: Question everything, and get down to detail; reduce all problems to their basic components; learn more and more about less and less.

Descartes's approach generated the first specialist scientific research, which, in turn, triggered the Industrial Revolution and, with it, Adam Smith's idea that output was increased if the different stages of production were divided among different workers.

Reductionism and the division of labor have given us the highest standard of living in history. They have also brought socially unsustainable rates of innovation and population growth, and the kind of specialist thinking that makes it difficult to see beyond the end of your Ph.D. As a result, commercial-secrecy-shrouded research labs, working on everything from new pesticides to smart bombs, launch their latest successes onto an unsuspecting market, and when these new products bump into other equally unexpected novelties—because of the way the world is networked—the result often causes unforeseen ripple reflects. For instance, Edison's electric light threatened the gaslight business, which was then given a temporary reprieve through the invention of the incandescent gas mantle by Auer von Welsbach, whose mantle research also revealed the existence of the rare earth neodymium, later available to dope the crystal for the first laser (itself fundamentally based on the electronic behavior of one of Edison's lightbulbs).

Everything is connected. As you read these words, somewhere someone you've never heard of is doing something that will sooner or later bring change to your life. And sometime in the course of the next twenty-four hours you'll do the same to others. None of us is untouched by the swirl and eddy of serendipity that drives human endeavors at all levels from quantum chromodynamics to painting your house.

In the past, the ripples took longer to spread because we were few and communication was slow. But the process was essentially the same as it is today. No decision, or course of action, escapes the effect of chance. For example, at the Battle of Hastings in 1066, if the English instead of the French had used the new stirrup to field mounted shock troops, and thus were victorious, this book would have been written in a form of English unaffected by the post-Hastings French invasion of England, and the text would look more like: *þa Frencyscan ahton wælstowe geweald* ("the French won").

Until recently, reductionism discouraged the cross-disciplinary, connective view of events because we lacked the means to gather and cross-refer the very large amount of data that would have made such an approach feasible. So we tended to organize history as we organized knowledge: in specialist terms, boxed into separate, straight-line, thematic structures. And yet the most cursory examination reveals this is not the way things happen. For instance, as this book shows, the emergence of stealth aircraft came not so much from earlier work in the field of aeronautics as from crystal-diffraction studies and audio recording tape technology. Above all, like everything, stealth technology was the end product of a series of human encounters, each one as accidental as the last.

The point of looking at history like this rather than in the traditional way (in terms of themes, or Great Moments, or leaders-who-showed-the-way) is because it offers a postreductionist *systems* approach to the turbulent modern world in terms of the whole rather than the parts. And because the more we recognize that we are all linked by encounter, from one end of the planet to the other, the better.

The reason I have straitjacketed the serendipity into twenty-five tales, each with a common beginning and a common end but different middles, is because there are intricate and fascinating patterns to be made out of the chaos of history.

I like the look of this particular pattern. I hope you do, too.

HOW TO READ THIS BOOK

Each chapter opens with a paragraph on the trigger event that kicks off the twin-track storylines.

Track One then runs, on successive *left-hand* pages, until: "End Track One."

At this point, don't turn the pages to see the chapter ending unless you're one of those people who likes to go to the back of the book and see who did the crime before you start the thriller.

Return to the beginning of the chapter and this time read Track Two, which runs only on the *right-hand* pages, until: "End Track Two."

Read the chapter ending.

Repeat as required, or until the onset of sleep.

ONE

1804: ATTACK ON TRIPOLI TO FISH STICKS

The first time the United States directly attacked Tripoli was at 9:47 P.M. on September 4, 1804. Under the watchful eye of the USS *Constitution,* the fireship USS *Intrepid,* packed with gunpowder and shells, sneaked into Tripoli harbor and blew itself up. This incursion was in response to four years of attacks by Tripoli pirates on American Mediterranean shipping, with the loss of one American ship and her three-hundred-person crew, at the time of the attack languishing in Tripoli jails (and, soon after, released).

The man controlling events that night, and in overall command (of the *Constitution,* three schooners, and eight other ships: a total of 156 guns and 1,060 sailors), was the bad-tempered Commodore Edward Preble, a veteran of the War of Independence. Preble had been ordered to make his base at Valetta on the island of Malta but, for various reasons, preferred Syracuse on the island of Sicily. Malta was British at the time, which might have had something to do with Preble's Sicily decision. At one point, Preble and his fellow officers dined with a visiting (and rather inquisitive) Brit, who, unknown to Preble, was working as spy and dispatch-writer for the Governor of Malta, Alexander Ball; Ball, a naval officer (and friend of Nelson), was an old hand at running ships and islands but less good at prose. The scribbler in question had left England for Malta for reasons of health and was, by this time, trying (and failing) to kick his opium habit, while continuing to pen the stuff that would make him one of the most famous of all Romantic poets: Samuel Taylor Coleridge. In 1805—having failed to give up drugs and yearning for the drizzle—Coleridge left Malta for London, via Rome, where he heard the news that Napoleon had him marked for assassination because of some earlier article he'd written in the *Morning Post.* The purveyor of this tidbit was the Prussian representative to the Vatican, Wilhelm von Humboldt.

By this point Wilhelm was a well-known esthete with some major literary criticism work behind him. He would also go on to become a lead player in comparative linguistics and fail to complete a great work on some obscure Javanese dialect. Prussian liberals like Wilhelm helped bring about teacher-training reforms and the establishment of a university in Berlin. They also talked a lot (cautiously) about civil rights and how the powers of the state should be limited. Most of this spirited chatter went on at the Berlin elite-meet salon (where Wilhelm dropped in from time to time) run by the extraordinary Rahel Varnhagen von Ense (née Levin), upwardly mobile daughter of a rich businessman. To her contemporaries, von Ense was the most cultured woman in Europe (only Mme de Staël might have disagreed). For a few years at the beginning of the century, von Ense organized gatherings that attracted princes, commoners, composers (Mendelssohn), thinkers (Goethe), poets (Heine), Jews and Christians, Germans and foreigners. You were welcome if you had a point of view, a witty tongue, or intellectually demanding matters to reveal. As was the case with the Reverend Friedrich Schleiermacher, a salon regular and local preacher. Schleiermacher was to religion what the Romantics were to the arts: a reaction to the rational excesses of the Enlighten-

TRACK TWO

On board the USS *Constitution* that night was Lieutenant Isaac Chauncey, who did so well during the Tripoli war he ended up in charge of all naval forces on Lakes Ontario and Erie. Where, from 1813, he ran the first proper arms race in American history, launching ships as fast as he could build them so as to clobber the Brits, who were on the Canadian lakeshore launching ships as fast as they could build them. Before a full trial of strength could happen, the War of 1812 ended. Chauncey's boss was William Jones, who was then invited to become acting treasury secretary. A year later (1814), the new economics czar resigned because his personal finances were in total chaos and he was up to his ears in debt. So, naturally enough, when a decision was made to set up the Second Bank of the United States, Jones was the first choice to be its president. Things at the new bank went rapidly down the drain and included allegations of fraud on the part of Jones. First off, the bank had expanded with dangerous rapidity, and then gave customers such easy terms that everybody and his dog borrowed money and speculation became rife. A month later, when things started going wrong, the bank recalled every loan. Property values dropped fivefold in some places, and all over the nation thousands of individuals and small businesses went bankrupt.

One such loser was would-be bird painter J. J. Audubon, whose Mississippi steamboat enterprise sank like a stone, taking with it the entire savings of a newly immigrant, newly wed couple named Mr. and Mrs. George Keats. Back in England, George's brother, poet John Keats (who'd lent them some of the money), went ballistic, vowing to clobber Audubon at first opportunity. Keats was the archetype of all Romantic poets: produced for only five intense years, was pale and wan, wrote about unrequited love and suicide and lovers' chopped-off heads, and caught the tuberculosis that killed him when still young. Quite apart from American money worries, Keats was always desperately short of cash. So when magazine proprietor and publisher John Taylor not only offered to print Keats's next epic offering, "Endymion," but also to come up with a healthy advance, Keats was as happy as a pig in manure. Taylor himself had, to this point, pursued an innocuous existence as a journalist, publisher, and writer on economic matters.

ment. He held that belief wasn't something to be objectively analyzed and dissected. *Au contraire.* It was a "mystical," utterly "subjective" experience that left the believer with a "feeling" of "absolute dependence." It was only though this immersion in the "sensation" of belief that one came to God. (If you read only what was in quotation marks, you've read key words from the Romantic Movement manifesto.)

In 1824 one of Schleiermacher's minor pieces (on the Gospel of St. Luke) was translated into English, and so impressed the ecclesiastical powers-that-were that it achieved for the translator the prestige job of bishop of St. David's in Wales. The high-flyer in question rejoiced in the anagram-fodder name of Connop Thirlwall. Began as a priest, then became a lawyer, then a classics don at Cambridge—where he made waves by saying that low-church Protestants should be let into the Church of England—was fired, became vicar of a church in bucolic nowhere, then finished his multivolume *History of Greece,* and was elevated to the episcopacy. Thirlwall's *History* was published by the then-famous Dionysius Lardner. Regarded as a major science popularizer (or charlatan, depending on who was regarding), Lardner forecast the link to India through the Red Sea long before the Suez Canal, and lobbied for transatlantic steamships when people thought the idea of dropping sail was crazy. It was during his early years as professor of natural philosophy and astronomy at London University that he began his great *Cabinet Cyclopedia* (133 volumes, edited over twenty years—the *Encarta* of its day). Contributors were legendary, including Charles Macintosh, Sir Walter Scott, Sismondi, and Herschel. Lardner also included a young writer trying to make money to support her child after her husband had been drowned in a sailing accident in Italy in 1822. Mary Shelley—author of *Frankenstein,* pal of Byron, daughter of feminist Mary Wollstonecraft, fancied by French novelist-antiquarian Prosper Mérimée, grieving and beautiful widow of tragic poet Percy Bysshe Shelley—was everything a Romantic was supposed to be. In her later years she scraped a living from writings, which included the piece for Lardner on Italian literature.

Mary dedicated her last effort (*Rambles in Germany and Italy,* 1844) to a longtime friend, one of those poets who sink almost without trace. Ever heard of Samuel Rogers's "The Pleasures of Memory," 1792? What Rogers lacked in talent he made up for in generosity. Having inherited a fortune at a young age, he proceeded (via a large and expensively decorated London house) to entertain anybody who wrote better poetry than he did. This had to be a great many and included Byron, Shelley, Wordsworth, and Sheridan. All his life, Rogers continued to churn out poetry so bad that only he would pub-

Then in 1859, out of nowhere, came his *The Great Pyramid: Why Was It Built?* Taylor was convinced the Giza pyramid wasn't Egyptian at all, but had been designed by an Israelite (maybe even Noah himself) acting under divine orders. Furthermore, Taylor opined, the numbers relating to the pyramid's complex dimensions hid a secret, encoded message of universal importance, from you-know-who. This claptrap proved to be irresistible to Charles Piazzi Smyth, who was otherwise totally sane. Smyth was an astronomer, Royal Society fellow, and pal of serious stargazers like Herschel. Nonetheless, bitten by the pyramid bug, at the height of his career he went off to Giza, measured every inch of the pyramid, and in 1865 announced that the "secret code" explained everything in the Old Testament and foretold the Second Coming. As a result of which the Royal Society booted him out.

But others took up the mystery. Was it a coincidence, they asked, that the pyramid "inch" was exactly the same as the Imperial British inch? This fatuous but "strangely convincing" load of hocus-pocus was given the coup de grâce in 1880 by the down-and-dirty, in-the-trenches work of archeologist hardhead Flinders Petrie, whose dad had been a Pyramidology convert. Petrie's opinion on the matter was expressed in a paper written after exhaustive measure-and-dig efforts, and called for archaeology to be more brush-and-scrape routine and less now-it-can-be-revealed gobbledygook. His opinion of Pyramidology can be summed up in one word: "garbage." Petrie set the tone for all later excavation, as he went through sites in Egypt and Palestine like a hot knife through butter (cut a trench, look at the layers, reveal the historical sequence). He was able to do this in Palestine, thanks to the energetic Palestine Exploration Fund money-raising capabilities of a great Victorian amateur, George Grove. Grove began as an engineer, working for the likes of shipbuilder Robert Napier and bridge builder Robert Stephenson, then graduated to secretary of the Society of Arts, music criticism and analysis, friendship with the musical greats, first director of the Royal College of Music, and finally, editor of the *Dictionary of Music* that now bears his name (and saves all of us long research hours in the library).

In 1915 Grove's granddaughter Stella proposed to Peter Eckersley, and they were married. Two years later, Peter joined the Royal Flying Corps as a wireless equipment officer. In 1922

lish it. Nonetheless, he must have impressed a senior bureaucrat because when Wordsworth died in 1850, Rogers was offered the poet laureateship. He declined the honor, so they gave it to somebody named Tennyson, by this time pulling out of a struggle with booze and with some very respectable work behind him. Tennyson was the Victorian poet par excellence, all gloom and saccharine. Apart from one foot wrong—written during the Crimean War ("The Charge of the Light Brigade" hinted at such incompetence in the army that it outraged every right-thinking harrumph)—Tennyson could do no wrong, especially after Queen Victoria gave him the ultimate nod. Throughout his writing career, Tennyson returned again and again to his love of the medieval and in particular King Arthur, with highly polished stuff like "Morte d'Arthur" and "The Lady of Shalott." Tennyson's knights-of-old flimflam fired the callow imaginations of every undergraduate, particularly those of William Morris and his pals Burne-Jones and Rossetti, who took medievalry further over the top, inventing the Pre-Raphaelite school of painting, and cutesy pseudo-fourteenth-century Arts and Crafts wooden furniture and flowered wallpaper that gave you eyestrain. All of which made them a fortune when it hit late Industrial Revolution consumers beginning to yearn for the imagined simplicity and tranquillity of a recently bygone age of prefactory country pleasures. This attempt to return to the purer life, of a time before the downtrodden proletariat existed, sprang from Morris's dyed-in-the-wool socialism.

This he shared with George Bernard Shaw, a down-at-heel, frayed-cuff, would-be journalist, who joined Morris's Socialist League in 1888. In 1893 Shaw caused a furor with his first (censored) play, about a prostitute. Pungent on-stage social comment followed in the shape of boffo successes like: *The Devil's Disciple, Major Barbara, Pygmalion* (in a later existence, *My Fair Lady),* and *Arms and the Man.* By the time he died at ninety, Shaw was considered the world's greatest living dramatist. Shaw socked it to all forms of what he considered humbug. Back in 1875 he wiped the floor with the visiting (and renowned) American evangelist Dwight Moody, after attending one of Moody's music-and-prayer revivalist meetings. Moody, who'd started life as a boot salesman, set the mold for revivalists thereafter: rugged physique, dark suit, homespun philosophy, plain ungrammatical language, and the message that God loved you no matter what.

This approach went over very big with a medical student, Wilfred Grenfell, who went on to became a medical missionary to deep-sea sailors. In 1892 he visited Labrador and was so shocked by the poverty that he stayed longer and set up the Labrador Mission. When he quit, forty years later, the Mission consisted, among other things, of six hospitals, seven nursing stations, four schools, a lumber-mill

he was working for the radio equipment company (founded by Marconi) to be given the first license for regular radio broadcasts, which Eckersley organized (and took part in as actor, announcer, stage manager, and engineer). For half an hour every Tuesday, his team filled the airwaves for those very few able to hear them. A year later, the monopolistic never-consult-the-listener British Broadcasting Corporation was founded, and Eckersley became chief engineer. His sidekick (assistant engineer) was Noel Ashbridge, who later rose to a position in which he made the crucial decision about which system ought to be chosen for the BBC's first TV broadcasts. He chose the twenty-five-frames-per-second, major-user-of-bandwidth, 405-line-scan approach pioneered by an extraordinary Russian immigrant named Isaac Schoenberg. The result, in London on November 2, 1936, was the world's first high-definition TV broadcast. Plaudits all round, and eventually (in the case of Schoenberg and Ashbridge) ennoblement as Sirs. Not so Eckersley, who was involved in a divorce and a whiff of scandal unacceptable to Auntie Beeb. Eckersley actually resigned (these were the days when standards were high and radio announcers wore evening dress). Schoenberg had earlier set up the first radio stations in Russia, before leaving in 1914 for pastures Western and more democratic.

Apart from Schoenberg's success with TV, he made another right move in 1929 when he hired a young engineer, Alan Blumlein, to develop a system which would save Schoenberg (and the Beeb) from having to pay through the nose for American sound-recording equipment royalties. Blumlein produced the required system, and then in 1931 filed the patent for a technique that would generate the kind of sound to be enjoyed when the listener was using more than one ear. In 1934 Blumlein recorded Beecham conducting Mozart, with a recording stylus vibrating in two directions (in response to two incoming signals): one vertical and the other (in the same groove) lateral. We call what Blumlein made possible "stereo." Stereo first hit the general public in 1940 with Walt Disney's *Fantasia,* recorded in stereo by the Philadelphia Orchestra and conducted by Leopold Stokowski, who believed Hollywood could bring good music to the masses. Initially, he was wrong. It would take until 1960 for the mix of Mickey Mouse and the music of Bach, Tchaikovsky, Dukas, Stravinsky, Beethoven, Ponchielli, Mussorgsky, and Schubert to become a cult hit. In the end, Stokowski's innovative approach (free breathing

cooperative, clothing distributors, and four hospital ships. In 1912 one of the temporary hospital-ship staff was a young man who had previously worked in the Labrador fur trade. He noticed that on days when the temperature was fifty below, whenever the local natives pulled fish out of the water, the catch instantly froze. And months later, when they thawed the fish out, he noticed that some of them showed signs of life. He tried the same trick on meat and veggies. All of which retained their taste and consistency if they were quick-frozen while still fresh. Could it be made to work on an industrial scale? Back in the States, by 1925 the young man was selling instantly frozen haddock fillets. After which, it was time for Clarence Birdseye to chill out and enjoy well-deserved fame and fortune.

END TRACK ONE

for the wind and free bowing for the strings, which produced the rich "Stokowski sound") and his penchant toward modern composers like Berg, Schoenberg, and such, made the Philadelphia old fogies see red, and Stokowski left for a flamboyant superstar life that included marriage to a Vanderbilt.

One of his pals was perhaps unexpected for someone so extravagant. Irving Langmuir was a self-effacing chemist (Nobel, 1932) whose research ran the gamut from ice crystals in clouds and floating seaweed orientation to smoke screens and (his main obsession) molecular and atomic structures. This included some original thinking about valence and bonding (the way in which atoms could share electrons). Langmuir's results encouraged chemists to approach the whole matter of how molecules happened in ways that turned out to have some interesting potential. At least it did for Thomas Midgley, working for a lab in Dayton, Ohio, and asked by his boss to solve the problem of knock (incomplete combustion in the cylinder, and no good for cars or drivers). Taking Langmuir's how-molecules-come-together approach to the elements, Midgley went through every single one of them, looking for molecular arrangement that might do what was needed.

Six years of minutiae later, in 1921 he found it: tetraethyl lead (the additive that gave gasoline the name "leaded"). Encouraged by this discovery, Midgley's boss then asked for a nontoxic, nonflammable refrigerant (those available at the time tended to leak and kill owners as they slept). When Midgley had it (this time it took him only three days), at the American Chemical Society meeting in 1930 he inhaled a lungful (to show it was nontoxic) and blew out a candle when he exhaled (to show it was nonflammable). The new wonder product became known as Freon. Ironic that decades later Midgley's wonder chemicals should turn out to be bad for the individual (lead poisoning) and bad for the planet (ozone hole).

END TRACK TWO

AND FINALLY . . .

Thus it was that when Clarence Birdseye's early fresh-frozen fillets were coming off his superchilled production line, Midgley's Freon-filled refrigerators were there to store them in. Fish sticks were here to stay.

TWO

1760: FAKE EPIC TO ORGAN TRANSPLANTS

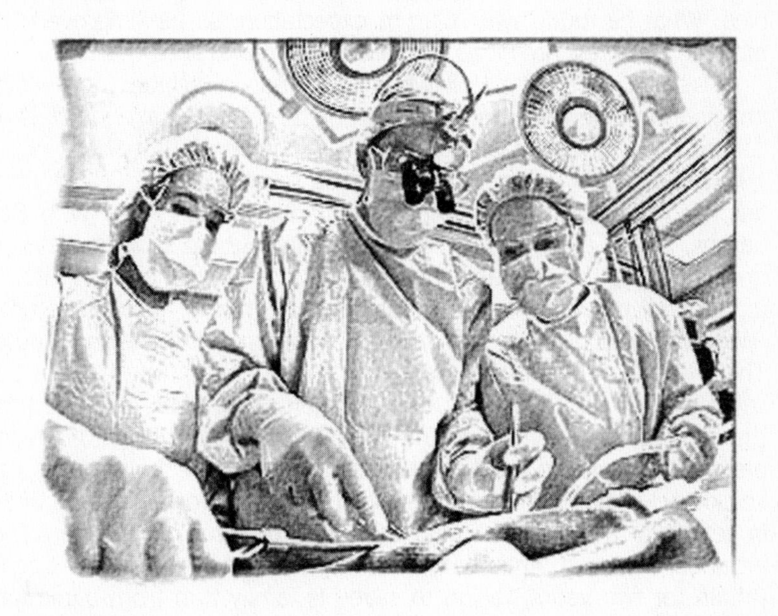

Few literary endeavors have succeeded in fooling major minds of the caliber of Napoleon, Coleridge, Mendelssohn, Goethe, Byron, Massenet, and even strait-laced economist guru Adam Smith. But *Ossian* did. What's more, in getting under the skin of German trend-setter J. G. Herder, *Ossian* could be said to have inspired Herder to kick off the entire Romantic movement. *Ossian* was a collection of fragments of epic verse (by the third-century Gaelic poet of the same name) collected by a Scots scholar, James Macpherson, and published in 1760. And *Ossian* was a fake.

James Macpherson was well intentioned. At the time, Scots culture did not look to be long for this world. After the abortive Highlander attempts to snatch the British throne in 1715 and 1745 (by father, then son, James and Charles Stuart), the English cracked down on the Scots with a vengeance. Declared illegal: carrying arms, holding clan meetings, wearing tartan, and speaking or doing anything Gaelic (like playing bagpipes). Macpherson would eventually spend time in Pensacola, Florida (as secretary to the governor), and then in England (as agent for the Indian Nabob of Arcot), after his trip to northern Scotland to collect folk tales before the Scots and their Gaelic-speaking performers died. What he found wasn't up to expectation, so he "improved " the material and presented it as the real McCoy (so to speak).

Ossian created the Hollywood Highlands of a later, non-Scottish imagination (see anachronisms such as the movie *Braveheart).* It also caught the fevered imagination of high schooler Walter Scott, who by 1802, had grown up and was doing the same trick. For fellow Victorians, Scott generated an entirely fictitious Middle Ages with Scots barons, Saxon earls, and Crusader knights. He did so with plays (not good), poems (so-so), and the first historical novels (smash hits). He did so well he was made a baron, and the title of one of his novels *(Waverley)* was given to the main train station in Edinburgh, Scotland. That's fame. With all the money he made (he churned out a bestseller a year), in 1822 Scott built a fake baronial hall (renamed from the infelicitous Cartyhole to the more appropriate Abbotsford)—complete with turreted tower, wainscoted rooms, echoing halls, and knickknacks like Rob Roy's purse and Lord Montrose's sword. Writers in search of the same success beat a path to his castle door. These included a young American, Washington Irving, who had been in Britain for two years, failing to save his family firm from bankruptcy, and who would stay on for over a decade, writing cute stuff about the English. As a result of which, upon his return to the United States he was offered nomination for the U.S. Congress, mayor of New York, and secretary of the navy. While in England, in 1822 he in turn had suffered the visiting-American syndrome, in the person of George Bancroft, heading home from a doctorate at Göttingen (first to an American) and a Kultur-tour of Europe, meeting the likes of Goethe and Humboldt. Now well-educated, Bancroft returned Stateside and embarked on what would become a ten-volume *History of the United States* plus an eleventh ("Author's Last Revision"). By the time it was finished, scholarship had, unfortunately, moved on. Meanwhile Bancroft had returned to Europe twice. Once as U.S. minister to Prussia (loved Bismarck) and, earlier, to Britain as ambassador.

TRACK TWO

The palace of Fingal, the hero of *Ossian,* was called Selma—which, in 1816, gave William Rufus King the name of the new town in Alabama he was founding at the time. King was in Congress for almost a record twenty-five urbane years. You needed ruffled feathers smoothed? King was your man. His other claims to fame were a relationship with President Buchanan, which may or may not have been sexual, and the fact that he failed twice to be nominated for vice president and was then third-time-unlucky, dying shortly after being sworn in. Back in 1816 (after Selma), King sailed for Naples, Italy, as American legation secretary to William Pinckney.

On the same ship was a young midshipman nobody, who would one day become the Admiral Farragut of "Damn the torpedoes!" fame. Farragut's on-board math tutor was one Charles Folsom, recent Harvard divinity graduate, who then spent five years in various Mediterranean libraries and as U.S. consul in Tunis, before returning home to become a librarian and book editor for the rest of his textually nit-picking life. Most authors hated his perfectionist approach to their sloppy style (don't we all). About the only excitement in his otherwise-placid existence came in 1825, when he briefly became co-editor on the *New York Review* with William Cullen Bryant. Bryant had begun as a New England poet, with meaningful stuff on death and nature, and by the time Folsom met him, was up there on Parnassus with the best of the belles-lettres American esthetes. In 1827 Bryant joined the *New York Evening Post* and two years later was editor in chief and co-owner. In tune with his own predilections, Bryant gave his readers a stimulating diet of no-holds-barred editorials on abolition of slavery, copyright, prison reform, the nonannexation of Texas, and the establishment of a large central park in New York. Eventually he persuaded John Bigelow to come aboard and share the load.

Bigelow matched Bryant's reformist views on almost everything. When the Civil War broke out, Bigelow was sent to Paris as consul general and became indispensable because the incumbent U.S. ambassador didn't speak French. While there, Bigelow almost (but not quite) persuaded the French and the English not to build any warships for the Confederate navy. He did well enough, how-

Britain is where Bancroft met and no doubt fell for (everybody did) the magnetic Angela Burdett-Coutts, the richest heiress in the country, second in public popularity only to Queen Victoria and philanthropist extraordinaire. She put most of her megafortune into reviving Irish fishing fleets and saving destitute children, orphan animals, schools, churches, African bishoprics, Brittany lifeboats, soup kitchens, hospitals, public housing, fallen women, and cotton gins for Nigeria, and above all, spreading British influence among the "savages." The latter attracted her to James Brooke, an attractive man with an attractive colonial idea. In 1847, when they met, he was already all over the newspapers for his spectacular and daredevil adventures with the headhunters of Sarawak, where he'd been made rajah by the local sultan, and then (with the help of the British navy) wiped out piracy in the area, introduced taxes, set up law courts, and generally started bringing the locals a whiff of British life. Brooke was the first of what would be three generations of "white rajahs of Sarawak," and Angela couldn't get enough of him and his mission. This meant lending him millions, buying him a steamship, lobbying Parliament to recognize him and his fiefdom (it did), and getting him a knighthood. In return he (temporarily) bequeathed her Sarawak in his will. Brooke was everybody's dashing hero—like the bluff-heroic Elizabethans in *Westward Ho!,* written by another fan of his, Charles Kingsley, who in 1855 dedicated the book to Brooke.

Kingsley was an overworked parish priest who made it to royal chaplain because of his social work (and in spite of his Christian Socialism). He believed that Rome was the Devil's spawn, that the Anglican church should get into social and political issues, that science and technology would bring good things to life, and that the workers had a raw deal. His 1863 novel, *Water Babies,* summed up what he cared about: the lives of the poor, public health, education, pollution, and Darwin's theory of evolution (published four years earlier). There was also just a hint of anti-Semitism in his slightly Ayran version of Christianity-untainted-by-degenerate-Judaism. In 1873 he failed to be elected secretary of the Royal Academy, in spite of impressive backing from people like Fred Furnivall. Furnvall was a Christian Socialist on steroids. He believed that trades unions were an essential part of Anglican theology. For years he taught at the Working Men's College in London. And although *your* reaction just now may have been "Fred who?" he lives on in the *Oxford English Dictionary,* of which, for twenty years, he was first editor. And from 1853 on, Furnivall set up more societies dedicated to reprints of original manuscripts than you could shake a quill at: the Early English Texts Society, Chaucer Society, Ballad Society, Shakespeare Society, Wycliffe Society, Browning Society, and Shelley Society. One of the

ever, to be promoted to ambassador in 1865. He'd landed his original Paris posting because he knew William Seward: lawyer, senator, New York governor, and deal-making politician. Seward was one of those people who knew where the bodies were buried, so he usually got what he wanted—usually behind tightly closed doors. This was how he handled the deal of all deals (and the one that put him in the history books). Seward was keen on international trade, international telegraphic communications to facilitate trade, an international currency to make trade easier, and U.S. expansion into the Pacific to foster trade. So when in the mid-1850s the initial failure to lay a transatlantic cable got people thinking about possible alternate routes (like one stretching from America across the Bering Strait, across Russia, and on into Europe), Seward led the charge—especially when he heard a rumor that the Russians might want to sell Alaska. He looked around for ammunition and found it in the person of Fullerton Baird, Secretary of the Smithsonian, who got his research people to turn up reports of rich Alaskan coal deposits, inexhaustible fish and seal supplies, and too few natives to cause a problem.

In fact, the Russians had wanted to get rid of Alaska for a long time. It cost more than it made. Besides, Russian reasoning was that if the faraway place became attractive enough to the Americans, there would be little St. Petersburg could to do forestall a U.S. invasion. Among themselves they agreed they'd accept five million dollars, but Seward was so keen the price went up to seven, and Seward didn't blink. The matter was managed by the crafty Baron Edward de Stoeckl, negotiator on the Russian side, who'd spent years in Washington, knew how these things were done, and very probably spread a little slush money around to make sure Seward got congressional support. Stoeckl was an Austro-Italian freelancer, and the Russians liked his negotiation skills so much they hiked his pay. And when Seward suggested the entire matter be kept secret till negotiations were over (1867) so Congress would have to agree a fait accompli (1868), Stoeckl magnanimously agreed, while writing home about dumb Americans.

Stoeckl was a schmoozer to his fingertips (an American wife helped). One night back in 1854, during some diplomatic glad-handing, he was hijacked by a young, oddly dressed draftsman from the U.S. Coast Survey, who took Stoeckl home, cooked dinner, and turned out to be a bril-

other people who had joined Furnivall in backing Kingsley's failed application was a French visitor, Hippolyte Taine, lecturing at Oxford on French literature, and an expert on *l'Angleterre* (he wrote: "Sunday in London presents the aspect of an immense, well-ordered cemetery"). Taine wrote prolifically on art, philosophy, and literature, and where Darwin came into them all. For Taine, "Vice and virtue are products like vitriol and sugar." Human culture was the product not of some intangible talent like imagination or caprice but of environment, race, and the contemporary context. The culture of each age was different because the circumstances of each age were different. And you could only understand artistic expression by studying it scientifically, the way you would study nature. Hence the name for Taine's way of thinking: "naturalism," which in its time was shocking and "impactful" in a way it is not now (as Taine himself would have said).

One of the many people impacted was Émile Zola, journalist and writer of too many novels to mention, except for the monumental twenty-book series with the general title: *The Rougon-Macquart Family: a Natural and Social History of a Family,* which put Taine's naturalist ideas into practice. From 1870 to 1893 (at a rate of just under one novel a year), Zola wrote about members of the same family over the period 1851 to 1871 and showed how their lives were conditioned by their changing environment and by heredity. As each novel hit the bookshops it generated extreme reactions: You either loved it or hated it. In England the novels were "modified" before being published to save readers the embarrassment of Zola's overrealistic language. In France the last novel of the series came out the same year that anarchist Auguste Vaillant threw a small bomb from the public gallery of the French Parliament, wounding many but killing none ("Had I wanted to kill I'd have used a gun"). Vaillant was sentenced to death, and Zola joined many journalists and public figures petitioning French President Sadi Carnot to commute the sentence. Carnot refused and on February 5, 1894, Vaillant was guillotined. On June 24 that year, President Carnot was on his way to the theater in Lyons when he was stabbed by an Italian anarchist, Santo Caserio, in an act of revenge for Vaillant's death. The knife severed a major artery, and since there was no way to repair such a wound, Carnot bled to death.

The event profoundly affected Alexis Carrel (a French surgeon working in a Lyons hospital that day), who decided to find a way to solve the problem. He first took embroidery lessons, then in 1902 developed a technique using an extremely thin needle and fine silk thread. The technique became known as triangulation, because it consisted of folding back both ends of the severed blood vessels, pulling both cross sections out to form a triangle, sewing a single joining stitch at the points, then stitching along the three edges.

liant and amusing host. The following year, James Whistler left for Europe. Unlike many young men who did the same, Whistler became a famous and successful painter. After the obligatory session in Paris, he got into his stride with the work for which he was to become known: *Harmony in Blue and Brown, Symphony in White, Nocturne: Grey and Gold,* and, the one for which he is remembered best of all, *Arrangement in Grey and Black* (a.k.a. *Whistler's Mother*).

Whistler's emphasis on color harmony rather than subject matter underwhelmed some people. People like Fleeming Jenkin, an engineer who spent most of his life laying submarine telegraphs (Red Sea, Brazil, Atlantic) and getting turned on by anything to do with cables—in pursuit of which he wrote the definitive paper on the specific induction of gutta-percha and classified the ohm as an absolute measure of resistance. Like a true Victorian, he also pontificated on geology, animal breeding, heredity, monorails, public health, and plumbing. In spite of all this, he was known for his jolly humor and engaging character (presumably you had to be there). In 1871, while he was Professor of Engineering in Edinburgh, a favorite student quit engineering for the law, in spite of which the two remained lifelong friends. Robert Louis Stevenson had decided to be a writer back when his grandfather knew Sir Walter Scott. But his was a lifelong struggle against ill health, which seemed to improve the farther south he went: London, Bournemouth, Nice, New York State, California, and finally Samoa. By 1882 he'd written *Treasure Island,* and his book was in every family library, with income to match. Success followed success, with *Black Arrow, Kidnapped, Dr. Jekyll and Mr. Hyde,* and *Master of Ballantrae.* In 1888 he and his American wife and stepson chartered a yacht and set out from San Francisco, discovered the delights of the South Pacific, and never returned.

While still in drafty Bournemouth, Stevenson was visited by an American expatriate, who was settling into Europe for life (and would eventually take out British citizenship) and who was about to produce a string of books that would open up the world of the psychological novel and play endless variations on the semiautobiographical theme: American hero comes to Europe, suffers culture shock, and asks self meaningful questions about the human condition. This, after Henry James had spent time meeting the demimonde of Paris, Rome, Florence, and London. By 1877 he

TRACK ONE

Blood flow could immediately be restored in the repaired vessel. In 1906 Carrel took up a post at the Rockefeller Institute in New York, where, with a few gaps, he worked for the rest of his life. After developing ways to transplant sections of undamaged blood vessels, for use in graft work, Carrel then moved on to tissue culture (with nutrient replenishment he kept a piece of chick heart tissue alive for several decades), and by the 1930s was ready for much bigger stuff.

END TRACK ONE

was well known enough to the elite to achieve honorary membership of the Athenaeum Club (fogies only) and receive 109 dinner invitations in one winter. He even did a stint in the country (Sussex), where the patrician American met a working-class, misfit socialist named H. G. Wells, who was busy trying to change the world. Wells was already working on the series of novels that would make his name as a futurist: *The Time Machine* (1895), *The Invisible Man* (1897), and *The War of the Worlds* (1898). In spite of the fact that the majority of his books were social and political in content (Wells believed in world unification, disarmament, and conservation), his accurate prediction of the atomic bomb and World War II has permanently associated him in the public mind with science (though he never got the Royal Society membership he so craved).

One of his earliest fans was a boy on a branch outside Boston. Having just read *War of the Worlds,* the lad was seized (up his aunt's apple tree at her farm in Auburn, Massachusetts) by the thought of flying to Mars. So after spending a few years getting an education in physics, he set about making such a thing possible. On March 16, 1926, he flew the world's first liquid-propellant rocket, in a field near the farm. Then, thanks to a phone call in 1929 offering help with his research, Robert Goddard spent the years between 1930 and 1942 working out every problem in spaceflight that would one day be faced by anything, or anybody, landing on the Red Planet. That offer-of-assistance phone call had come from Charles Lindbergh, already famous for his transatlantic flight two years earlier. Congressional Medal of Honor winner Lindbergh would, however, end up on the cover of *Time* magazine for something far removed from both aviation and spaceflight. Between 1931 and 1935 he developed an artificial heart.

END TRACK TWO

AND FINALLY . . .

Lindbergh's artificial heart took the form of a sterile glass perfusion pump. It would circulate the nutrients which Carrel's organs required to stay alive out of the body. Between them, the two men made possible the techniques that would lead to the first organ transplants.

THREE

1805: BATTLE OF TRAFALGAR TO LASER

Just before the start of the Battle of Trafalgar (noon till 4:30, October 21, 1805, off the Iberian coast), in which Napoleon's fleet would be clobbered and Admiral Lord Horatio Nelson (hero and lover) would cash in his chips, one of history's more famous pieces of editing occurred. In Nelson's hurriedly planned signal "England confides that every man will do his duty," the word "confides" was not in the codebook, and spelling it letter by letter was going to take more time than Nelson had before the fight. On the advice of his flag-lieutenant the word "expects" (a word already in the codebook) was substituted.

TRACK ONE

The Trafalgar rewrite man was John Pasco, who'd been in the navy since the age of nine and with Nelson for the previous two years. Alas for Pasco, British naval tradition had it that after a victory (such as Trafalgar) the Admiral in charge always saw his officers right, job-wise. So Nelson's death during the battle put a bit of a crimp on Pasco's career. The best he could get, four years later, was the command of the HMS *Hindostan,* escorting a supply ship out to Australia. His on-board surgeon was a man with a botanic bent, Joseph Arnold, who spent most of the en route stopovers collecting plants. On his way back to England, Arnold's ship (and all his possessions) burned in Jakarta harbor. Fortunately a school pal was secretary to the lieutenant governor of Java and got Arnold free lodgings and another ship home. In 1818 Arnold returned East, to Sumatra, as botanist to the ex-lieutenant governor, now governor, of Sumatra. Shortly thereafter Arnold and the governor (and the governor's wife) trekked off up the jungle to look around and generally do their colonial administrative thing. During the walk, two things happened: Arnold contracted the fever that would, within weeks, kill him, and he and the governor found the world's biggest flower. Three feet across, weighing a colossal fifteen pounds, it was later named after both men: *Rafflesia arnoldi.* Stamford Raffles was your classic colonial Brit eccentric overachiever. With a fast-track promotion in the British East India Company (the private firm administrating all British colonial property out East), Raffles did unheard-of things like learning the local lingo and mixing with the natives, as well as spending all his time traveling round the property and meeting (and listening to) the inhabitants. He also did all those things we now criticize (reorganized agriculture in the Western mold, ditto to the local law). And bought Singapore for the Brits. On visits home, Raffles became pals with royalty (and more important, science supremo Joseph Banks), became a Sir, lost all his gigantic botanical collection in another one of those ship fires, and eventually—thanks to Banks—became the first boss of the London Zoo and wrote on Javanese natural history. It turns out that he got most of his tropical know-how from a guy he'd inherited from the previous Dutch owners of Java, the Pennsylvania-born Thomas Horsfield. Horsfield was merrily botanizing when the Brits took over, and Raffles kept him on, with the words noodlers love to hear: "Just let me know what you need." Horsfield did so, and as a result in 1819 sailed for London with a collection of two thousand specimens (no fire, this time), and the job of curator in the East India Company museum. A happy bunny—less so if he had known how long it would take the great Robert Brown *not*

TRACK TWO

Lieutenant William Standway Parkinson had been with Nelson since the earlier knock-down-drag-out bash with the French (a.k.a. the Battle of the Nile) in 1798. A year later, Parkinson's great career move was to hang an Italian admiral for treason, by order of Nelson, under very questionable circumstances (case opens at 9 A.M., sentencing at noon, hanging at 5 P.M.). Parkinson's promotion—at the recommendation of Nelson, whose recommendations on anything were close to Holy Writ—followed a few weeks later. After which, in 1800 Parkinson went home to marry the sister of the Reverend Edward Clarke of Jesus College, Cambridge. Clarke himself wasn't there. He'd been gone for a year on a trip, after one of his students came into a fortune and could afford to pay for both of them. Following adventures galore in Norway, Finland, Russia, Palestine, Greece, Egypt, and Cyprus, the pair returned in 1802—well, almost, given that arrival took the form of a shipwreck on the shores of southern England, during which they lost a lot of their 150 crates (of anything they had taken a fancy to, from a two-ton Greek statue to England's first kohlrabi).

Their companion as far as Scandinavia had been a colleague of Clarke's (another clergyman) Robert Malthus, who was off to the Frozen North to collect demographic data to back up his new theory of population, published in 1798 and already causing major teeth-sucking among fellow intellectuals. Malthus held that Armageddon was just around the corner, because thanks to the Industrial Revolution the population was rocketing. And population increased, he said, geometrically, while food supply only increased arithmetically. It didn't take a genius to see that starvation and anarchy were just around the corner and that, apart from "self-control" (a euphemism for abstinence), there was little to be done about it. One Swiss economist thought otherwise. Simonde de Sismondi (reckoned by the great Keynes to be hot stuff) was one of the first to draw attention to the appalling conditions in which factory workers lived and worked. But there was no need to worry since you could handle Malthus's little starvation-and-anarchy problem with government intervention in the form of unemployment and sickness benefits, minimum wage, profit-sharing, pensions, and other such loony socialist stuff. In an era when capitalism was really getting into gear, this was a red flag, literally (small wonder then that so many of Sismondi's ideas were "borrowed,"

to catalog and name all those specimens he had so carefully packed and transported.

Botanic megastar Brown was too busy and, besides, took eleven-week annual holidays and, besides, was all caught up in discovering things like Brownian motion (tiny particles in fluid move around in random fashion), getting medals for investigating mysteries like vegetable impregnation, and spending years handling the transfer of the collection of his friend Sir Joseph Banks to the new Botanical Department of the British Museum (the deal was that Brown became its curator). On a trip to Paris in 1816, Brown met the man whose work he would describe as "declamatory nonsense." In return, Alexander von Humboldt called Brown "the Prince of Botany" and remained an admirer for the next forty years of meetings and letters. Humboldt himself was already on the science and exploration front page following his recent five-year South American extravaganza. In 1829 he was invited by the Czar of Russia on the closest the great Humboldt would ever come to a junket: an all-expenses-paid, nine-month, nine-thousand-mile trip through Siberia as far as the Chinese border. He did it with his usual thoroughness: a temperature-geologic-mineralogic-biologic assay of everything and everyplace—and correctly predicted that diamonds would be found in the Urals. Humboldt also persuaded the czar to give the OK to a chain of magnetic observation stations. Geomagnetism was the new Big Science of the period.

It was not surprising, therefore, that Humboldt helped write the magnetic observations' instructions for the British 1839 *Erebus* and *Terror* Antarctic expedition, commanded by James Clark Ross, in search of the southern magnetic pole. Ross had discovered the northern one on another intrepid exploration of the wastes around Baffin Island with his uncle, back in 1830, and now the German science genius Friedrich Gauss was predicting the southern version would be at 66° south latitude 146° east longitude. For the magnetic community, this was nail-biting stuff. Apart from taking the day off on Sunday, September 25, 1841, in Tasmania—and thus missing a great magnetic storm (a row with Humboldt followed about outer space not caring about Church of England observances)—Ross did pretty well. The magnetic pole was not where expected but at 75° south latitude 154° east longitude. One of Ross's other pretrip advisers had been the German Christian Ehrenberg, also a pal of Humboldt's (they'd done Siberia together). Ehrenberg's earliest big adventure had been an Egypt-Red Sea trip in 1820, when he'd collected thirty-four thousand animal and forty-six thousand plant species (the expedition can't have left much behind in Egypt), during which Christian had spread himself thin, disciplinarily speaking: from coral polyps and windstorm

decades later, by Marx). When not expounding on social matters (most of the time), Sismondi also wrote literary criticism, in 1819 producing a modest work on southern European literature.

This knocked the socks off a young Russian poet (almost unknown outside Russia), Konstantin Batiushkov, who was nuts for the Renaissance poets and used Sismondi's stuff to prep for a history of Italian literature he was never to write. In the meantime, he turned out reams of fairly turgid Romantic verse of his own and managed a quite successful civil-service career. At one point in 1818 he talked the czar into sending him to Paradise (a.k.a. Italy), where inexplicably he became depressed and was back home within the year. In the midst of all this, a minor annoyance (the invasion of Russia by Napoleon) got him drafted into the army and sent west. On this single occasion, army life revealed itself to be fun and cultural, since Konstantin became adjutant to a family friend, the big cheese General Nickolay Raevsky, hero of the Napoleonic campaign (and who eventually paraded Russian troops down the Champs-Élysées to rub the French noses in it). Raevsky was a career officer, and although later on he got mixed up with a failed coup d'état, unlike many others he survived to old age. At one point, while taking his family to a health spa in the Caucasus and passing through some middle-of-nowhere spot, he came across a young, loud-mouthed pal of his son's named Pushkin. A poor aristocrat ("poor" still included nursemaids and French tutors), Pushkin had written plays and poetry criticizing the czar (when the latter had the power of the thumbscrew and much worse) and was lucky enough only to suffer exile to the Bessarabian sticks. From there he moved on to the Crimea, where the Raevskys had rented a house and where Raevsky's daughters taught Pushkin English so he could read Byron and become the Russian equivalent (though some say of Shakespeare rather than of Lord B.).

Before his ouster from St. Petersburg, back in 1819 Pushkin went to a concert and heard the Irish pianist John Field play, and the two of them became fast friends. Field had left England in 1802 after an apprenticeship to Clementi and recognition by Haydn as a virtuoso, and was to spend twenty-eight years in Russia, lighting cigars with his fees, drinking to excess, and enjoying the general debauchery that went with success. Meanwhile Field produced works (now ignored) that

dust to electric rays. He also zapped organisms with the new electric force to see what they did. But his real love was the very small and the very old (as in micropaleontology, of which he was German pioneer). At one point, his pal Humboldt pulled strings and got Ehrenberg a professorship at the University of Berlin, where he taught such eager beavers as Othniel Marsh, a young American stones-and-bones person with a yen to be professor of it, back at Yale, where his rich uncle, George Peabody, was funding a new natural history museum.

Meanwhile nephew Othniel wrote papers on the Nova Scotia gold fields and went on digs that identified over eighty new types of dinosaur. Uncle Peabody's $100,000 endowment got invested, and by 1876 had grown to $176,000. The financial genius behind this was a pal of Othniel's, George Brush, metallurgist at Yale and a good man with shares. George was a would-be farmer, who early on (1853) had been sidetracked into two chemistry trips to Europe, where the bright lights of the British Royal School of Mines and Liebig's hot, new German chemistry lab diverted him. So the nearest he ever got to sod-busting was picking up rock samples. He knew more about the minerals of Branchville, Connecticut, than anybody. He got the mineralogy bug as a grad student assisting Professor J. Lawrence Smith (another Liebig pupil) at the University of Virginia. The two of them clarified certain obscure matters regarding certain obscure American minerals. Enough said. Smith's output was nothing if not varied: he discovered coal in Turkey; invented an upside-down microscope; was definitive on phosphate marls around Charleston, South Carolina; and became the last word on meteorites. Sometime in 1877, he sent some samples to a French chemist. Paul Émile Lecoq de Boisbaudran was that rara avis in credential-ridden nineteenth-century France, a self-educated man. He started out working for his family's wine business, reading up on chemistry and geology in his spare time. Went nowhere academically; met nobody academic.

Despite all this, Lecoq de Boisbaudran became good enough at spectroscopic analysis (burn it, look at flames through a prism, and see black lines in the flame spectrum indicating what the burning stuff contains) to discover gallium, and things about certain minerals of which few have ever heard (including dysprosium, terbium, gadolinium). At one point, Lecoq de Boisbaudran was doing things to rare earths, especially one named didymium (are you still awake?). This was the sample Smith had originally provided. Didymium was thought to be a pure rare earth, but Lecoq de Boisbaudran had his doubts. The arcane dispute was settled some time in 1885 by an Austrian, otherwise busy trying to help his gaslight-company shareholder pals fight off the emerging challenge of the electric lightbulb. This he did by discovering that if you soak a cotton gauze in rare earths (tho-

back then made a major impact with the cognoscenti, as did his wonderfully expressive style of playing (drunk or sober). One of his devotees was Friedrich Wieck, a piano teacher in Leipzig who also ran a music lending library, sold instruments, taught theory, organized regular private concerts, and wrote music reviews. Wieck's claim to fame in the modern world is that he was also the father of a daughter named Clara, who (after taking her father to court over it) married the unstable, jump-off-bridge Romantic composer Robert Schumann, at the time a lodger and pupil in the Wieck household. Back then Clara Schumann was far better known than hubby Robert. By the age of sixteen, she was already an international celebrity, with admirers including Goethe, Mendelssohn, Paganini, and later (lovesick for her) Brahms. In 1863 she went to Baden-Baden to be near her pal, opera diva Pauline Viardot, and to perform at Viardot's little soirées for such as the empress of France, the king of Prussia, et al. It was at Baden that she met a Russian by the name of Turgenev, a scribbler lovesick for Viardot.

Like most of his fellow-countrymen writers, Turgenev had several run-ins with the czarist police for saying things he shouldn't (like describing the indescribable peasant living conditions in the countryside) and had therefore enjoyed his share of uncomfortable exile in the boonies—nothing so painful, however, as his unrequited love for Viardot. Turgenev had it so bad he gave up fame and fortune in order to live with the Viardots in Paris, until finally he gave up hope and died. Before this, however, he managed a few forays into town, where he met yet another author whose name was on the police blotter. The novel that had lifted the "radical, dangerous" Gustave Flaubert to fame, *Madame Bovary,* was described by critics (who didn't have to live with the consequences) as "realist," i.e., too outspoken for safety. The censor's overreaction made Flaubert instantly famous. One of Flaubert's pals, also deeply into realism, was professor of esthetics Hippolyte Taine, the man who first brought reductionist method to the arts. Taine's argument was that a proper appreciation of the arts required a scientific approach. Observation and experiment were the secret to understanding a work of art because such a technique would reveal everything about the physical and historical context from which the work had sprung. This kind of egghead stuff was so far over the head of the official censor, Taine got away without so much as a tsk-tsk. The only

TRACK ONE

rium and cerium) it'd go blindingly incandescent when burned. Hence the gas mantle, for people who couldn't afford the new lightbulb or who still only had a gas supply. He gave his shareholder pals at least a decade more of income. Come to think of it, the mantles are still in use, lighting up evening campers. The gas mantle made Carl Auer von Welsbach a superstar. The Austrian throne dubbed him a baron, and with brilliant originality, he chose as the motto for his coat of arms: "More Light." On his way to the right rare-earth mix for the mantle (during the mind-numbing, repetitive, detailed work—over hundreds of hours—which is so much a part of the scientific endeavor and of writing about it), Welsbach found that didymium wasn't pure but could be broken down into two distinct substances. He named them praseodymium and neodymium—one of which affects you.

END TRACK ONE

aggravation he got was from fellow French academics who thought he was talking unadulterated piffle. Foreigners were more favorably impressed.

In 1869 a young Dutch chemistry student, Jacobus van't Hoff, devoured Taine's work and took it to its logical conclusion, so to speak. If science were essential to appreciating the arts, was imagination the key to scientific advance? Van't Hoff went on to prove this point with imaginative new discoveries like asymmetric carbon atoms (a.k.a. 3-D views of molecules, known as stereochemistry). These efforts (for which he got the first chemistry Nobel) were enthusiastically promoted by a fellow chemist, Brit William Ramsay, who translated something similarly imaginative by van't Hoff on osmotic pressure. This, before going on himself to discover the inert gases in the atmosphere, by dint of experiments that sometimes involved imaginatively boiling off twenty tons of liquid air and examining what came out. Among other things, it turned out to be entirely new gases such as krypton, helium, argon, and (in 1898) xenon. This was lightbulb-flash genius at its best, as evidenced by what happened next with xenon.

In 1927 a young engineer, Harold Edgerton, working at the General Electric plant in Schenectady, New York, was wondering how to deal with giant electricity-generating turbines that were on the blink but spinning too fast to see what was going wrong. At one point, a sputtering mercury arc light he was using "froze" the spinning turbine blades for a split second. Bingo: stroboscopy. The question became how split-second could he get? In 1934 Edgerton got an electric spark in a glass tube full of xenon to make a flash lasting just one millionth of a second.

END TRACK TWO

AND FINALLY . . .

If you dope an yttrium-aluminum garnet rod with neodymium, you have a crystalline environment whose atoms will absorb lots of white-light radiation and then emit brief bursts of massively intense, coherent (doesn't spread) light, which is so powerful it'll cut metal or get to the Moon. You provide the required large dose of white-light radiation input with a xenon flash-lamp. The entire process is known as light amplification by stimulated emission of radiation, or laser for short.

FOUR

1726: ENCYCLOPEDIA TO VITAMINS

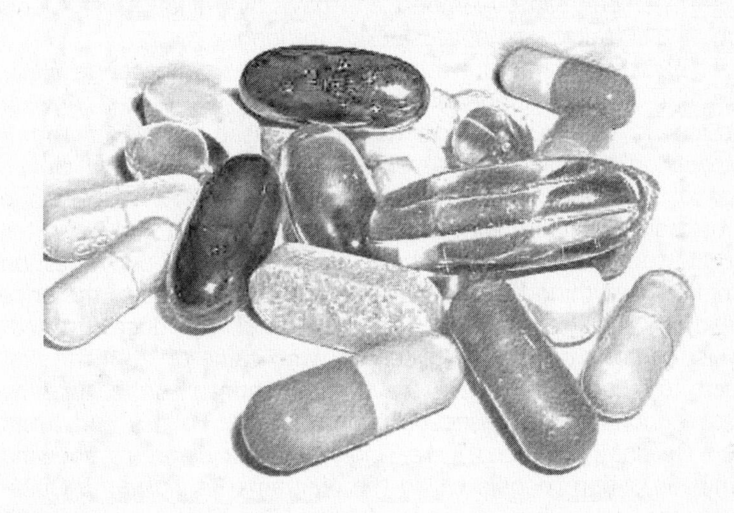

People in the early eighteenth century (those few who could read) felt, much as we do today, that the explosion of data and advances in communications technology put them on the threshold of a new and confusing age. The previous century had discovered ancient cultures in the East and produced an avalanche of never-before-seen stuff from the New World as well as amazing scientific novelties, including the vacuum (and all that came from it), the human circulatory system, the planets of the solar system, a strange new force named electricity, and too many others to name. And that was the problem. Things were beginning to move too fast for the average well-informed person. What was needed was easier, simpler, and better-signposted access to the vast and fissioning body of knowledge. Hence Ephraim Chambers's *Cyclopaedia* of 1726.

Ephraim Chambers's modest aim was to make all knowledge available in one digestible piece. So he organized his book alphabetically, rather than by subject, put a "map of knowledge" and a "chain of references" at the front, and divided everything into three sections (by sense, by imagination, and by reason) and forty-seven science and arts subsections. In 1734 Chambers's work looked like money-spinning to a French publisher, André Le Breton. But after wrangling with translators, Le Breton ditched the idea in favor of a new and improved (and all-French) alternative. For this job he hired the eminent math person Jean d'Alembert and gave him a journalist, Denis Diderot, as a sidekick—but not a sidekick for long.

In 1759 Diderot came up with a concept that would change Western thought: the *Encyclopédie*. As the apotheosis of the Age of Enlightenment, the *Encyclopédie* would be based on reason rather than revelation or belief. The seventeen volumes of text and eleven volumes of pictures Diderot finally published in 1772 employed a group of radical thinkers with a common aim: a giant show-and-tell work that would undermine Catholic authoritarianism and question everything. "Everything," as in: from the existence of God to the price of bread. This latter subject was discussed by doctor-turned-economist contributor François Quesnay, who came up with the first approach to economics based on scientific principles because he believed economics was based on a natural law. To this end, from 1758 on Quesnay produced increasingly complex diagrams showing economics in action (people called the diagrams *"le zig-zag").* Quesnay was the guy who invented *laissez-faire* ("don't mess with things"), the basis of all conservative planning since that time. Quesnay made it to the top because he obtained lodgings halfway up the stairs to Madame de Pompadour's apartments when she was Louis XV's live-in lover. Quesnay cured her frigidity (not good in a mistress) by weaning her off her usual diet of truffles, vanilla, and celery. Pompadour (real name: Jeanne-Antoinette Poisson) was probably Louis's nicest bedfellow. She was even a friend of the queen. A great supporter of all things cultural (Voltaire's patron), Pompadour had a particular thing about porcelain. When Louis gave her the town of Sèvres (near her château), she moved a porcelain factory there and started production. Above all, she loved doing makeovers to grand houses and then selling them. When her brother was made superintendent of the king's buildings, in 1750 brother and sister designed and constructed the École Militaire ("to show Louis' soldiers he loved them").

Nineteen years later, a Pole turned up on an army scholarship. Tadeusz Kósciuszko was a young military engineer, and while study-

TRACK TWO

The *Cyclopaedia* book of facts did so well that, in 1746, another publisher commissioned a book of words from a young man who'd arrived penniless in London only seven years earlier and had become an instant scribbler, bon vivant, and member (some say, inventor) of the chattering classes. Samuel Johnson took Ephraim Chambers as his model and over the next nine years scoured English literature to write the first modern lexicon—complete with the first standardized spelling—with precise definitions and quotations to clarify meaning where necessary (most famous: "Oats: a grain, which in England is generally given to horses, in Scotland supports people."). One of his drinking buddies at the Literary Club (Johnson was a depressive and staved it off with alcohol and conversation) was Richard Sheridan, playwright and canny politician who became liberal as, and when, necessary. As in 1799, when he joined the headline-grabbing campaign being run by well-known pious wet blanket William Wilberforce (who kept his opium addiction and early gambling excesses extremely quiet).

Wilberforce, one of the more ponderous members of Parliament, set up the Proclamation Society against Vice and Immorality. He also tried in vain for years to get Parliament to pass a bill to abolish slavery—until, in 1797, he added the word "gradual" to "abolition," after which things in Parliament began to move along more speedily, culminating in the act outlawing slavery in 1807. By 1805 Wilberforce was, therefore, the kind of luminary whom visitors to England would want to meet. Such a visitor was Benjamin Silliman, who'd just started the first chemistry classes at Yale and who'd come over to meet several science greats in the United Kingdom (Davy, Priestley, Dalton, etc.) and to travel around studying the latest high-tech marvel: canals. Back home, Silliman had been persuaded by the president of Yale to give up law and to put chemistry on the college curriculum. So he first went off to the University of Pennsylvania to get himself qualified and then taught the subject at Yale for fifty years. Shortly after his return from England, Silliman also started teaching the geology theory he'd picked up on his European trip, and then made the front page with his detailed description of the Great Weston Meteor, which fell on Connecticut in 1807. His analysis of the meteorite material (convincingly

ing in Paris, he met Ben Franklin and volunteered for you-know-what. From 1776 to 1783, he became a great American hero. Humane to his prisoners, fearless on the battlefield, sharing discomfort with his men, his imprint on America was the fortification he built at West Point and the suggestion he made for a military academy. Back home, he moved in and out of Poland trying to rid the country of the Russians—and failing, when he lost a decisive battle to Aleksandr Suvorov, one of those mythical Russian generals who fought in their underwear and slept on the ground. Suvorov led a charmed life until 1799, when he got to the Alps, where he was supposed to join forces with fellow general Rimsky-Korsakov and crush Napoleon's army. Only he found Rimsky-Korsakov had already been crushed by Napoleon's General Masséna, who led a charmed life of no defeats and whose career rose with Napoleon—from sergeant to marshal (1804), to duke (1809), to prince (1810)—thanks to his success rate. Masséna was probably Napoleon's best tactician. In 1799 his ordnance officer back in Paris was Pierre Daru, a consummate bureaucrat (and second-rate poet and historian) who looked after Massena's supply problems. Another Napoleon symbiont, Daru was councillor of state in 1803, count of the empire in 1809, and minister of war in 1811. Unlike many others, he survived Napoleon's downfall by disappearing into the country and coming back to pick up his pieces only when the dust had settled.

While minister of war, he looked after his cousin, Marie-Henri Beyle, for whom (this was during Napoleon's heyday) he fixed a job running supplies for the French occupation troops in the state of Brunswick, Germany. In that state, there was a small town whose name became the nom de plume by which Bayle is now remembered: Stendhal. After a shaky start, he did quite well with travel books like *Rome, Naples, and Florence* (1817). An unrequited affair triggered *On Love* (1820). Moving between France and Italy, Stendhal first hit the headlines with *The Red and the Black* (about a young peasant who seduces his way to success) and then *The Charterhouse of Parma* (about an aristocratic Italian family). "Acute psychological observation" is what the critics said. Balzac thought he was a genius. Late in life, on a boat trip down the Rhine, Stendhal met a couple of literary lovers. She (George Sand, novelist) didn't like him; he (Alfred de Musset, poet and playwright) did. When Musset had met Sand, back in 1833, Musset was a cause célèbre with his first poem *Mardoche.* Women sent him love letters. Now it was his turn to be smitten. He and Sand enjoyed five years, during which he wrote his best poems. Then she left with yet another lover.

Back in Paris, having fun with a bevy of talent like Delacroix, Ingres, Rossini, Liszt, and Chopin (another of Sand's lovers), one

proving that it was unearthly in origin) involved melting and analyzing various bits of it, using the oxyhydrogen blowpipe invented by Robert Hare, whom he'd met during his days at the University of Pennsylvania.

Hare was another example of the Great American Amateur. With only an honorary medical degree from Yale, he ended up professor of natural sciences at William and Mary and then professor of chemistry at the University of Pennsylvania. Hare invented many devices, including items for measuring gas density, specific gravity, and exploding gases, as well as a special "spiritoscope" (for preventing fraud by mediums in trance *and* to communicate with the dead). His original blowtorch helped create a new word in the English language, thanks to its use by Thomas Drummond, an army engineer who in 1819 was employed on the National Ordnance Survey of Britain. Drummond found British weather unhelpful (too murky) when trying to get a bearing on some distant triangulation spot. He used Hare's oxyhydrogen arrangement to direct a jet of flame onto a ball of lime with a concave mirror set behind it. The lime became incandescent, and the mirror directed the "lime" light. You could see it shine, in filthy weather, sixty-seven miles away—as Drummond found the first time he used it on location in Northern Ireland. Later the light was used in theaters to produce lighting effects and to illuminate the downstage position where only the lead actors got to stand (hence being "in the limelight"). In 1825 Drummond's first demonstration for the government was a knockout success. In lighthouses, it was clearly going to be clearly visible, from far away, by the growing number of ships that were hitting rocks instead of getting to their destination.

The limelight demo was witnessed by a navy captain, Basil Hall, who was no slouch at science himself. His claim to fame was landing on St. Helena in the mid-Atlantic, on the way back from a trip out East, and meeting the exiled Napoleon (who knew Hall's father and therefore made an exception to his never-see-anybody rule). After which, in 1820, Hall was posted to the River Plate and served under Commodore Thomas Masterman Hardy, who had a tricky time waving the British flag and being nice to all, during nonstop revolutions all over South America. Hardy had the dubious distinction of having served with British national role model Horatio Nelson for thirteen years and of having been next to him when the admiral got fatally shot on the

night Musset went to the opera and heard the outstanding, three-octave voice of a Spanish soprano, Pauline Viardot, who was causing a sensation. Viardot was extraordinarily talented both as a singer and as a composer. In 1843 in St. Petersburg, she met the Russian novelist Turgenev, who lived with her and her husband and children for the rest of his life. She worked with Gounod, Meyerbeer, Berlioz, and Chopin and wrote songs to the poems of Goethe, Turgenev, and Musset. After retiring from the stage at forty-two, she taught some of the great singers of the time, using methods developed by her father Manuel Garcia, a popular tenor who sang and directed operas in Spain, Italy, France, Britain, the United States, and Mexico. His son (Manuel) picked up on his father's teaching techniques and taught people such as the international soprano sensation Jenny Lind. However, the son really took his place in history in 1855, when, after studying the physiology of the voice (in French military hospitals), he invented the laryngoscope. One of his singing pupils was the Irish soprano Catherine Hayes, a singer good enough for La Scala, who had a moderately successful European career and then five years singing around the United States, the Sandwich Islands, Australia, Indonesia, and India.

The Irish were naturally proud of their megastar, so when she seemed to be libeled in a newspaper article, the Dublin public were up in arms; one groupie went so far as to take rooms next to the author and threaten a beating. Of course, it was all a mistake, as Thackeray (the alleged libeler) explained. He'd been talking about a seventeenth-century murderess of the same name. This happened while Thackerary was still on the way up, scribbling articles, reviews, sketches, travel books, and a column in *Punch* magazine, and before he'd written *Barry Lyndon* and *Vanity Fair* and other lengthy things. Thackeray was six feet three inches, had an enormous head, mixed with the London literary crowd, and (lonely and with a wife gone crazy) was a great clubman. Apart from the Garrick and the Reform, he belonged to one modestly named "Our Club," where he spent evenings with doctors like Benjamin Richardson. Richardson must have been fun, since he was rabidly antialcohol and one of the first public-health and bicycling obsessives. He then became known for inventing fourteen different kinds of anesthetic. Richardson's claim to fame rests with the development, in 1867, of a device that sprayed ether onto the skin and numbed it in preparation for some local scalpel work.

This spray caught the attention of Joseph Lister (at that time professor of surgery at Glasgow University), who was busy turning operations into survivable events with carbolic acid—first doused on dressings, and then (after seeing Richardson's spray) filling operating

Victory (Hardy was the ship's captain) during the Battle of Trafalgar.

Hardy's only other distinction was that he was a distant relative of Thomas Hardy, the quintessential BBC-TV-classic-drama-adaptation novelist, who lived in the late nineteenth century in deepest English bucolia and wrote nostalgically about a vanished world of agricultural towns and villages and the socioeconomic and sexual problems with which they were ridden (e.g., *Far from the Madding Crowd, Tess of the d'Urbervilles,* and such). Hardy was already A Great Novelist by midcareer, though his start was slow. The first three novels bombed (including *Under the Greenwood Tree),* much to the chagrin of his (immediately ex-) publisher William Tinsley. Back then, most novels were serialized in magazines, and Tinsley's magazine included pieces from names like George Meredith and Anthony Trollope, as well as unknowns like W. S. Gilbert, a young lawyer just breaking into literature. By 1861 Gilbert was contributing columns to *Punch* magazine, publishing the lighthearted *Bab Ballads,* and from 1866 writing plays which have all sunk without a trace. In 1875 everything for Gilbert changed with a show called *Trial by Jury.* Gilbert did the words; Arthur Sullivan did the music. Success came just in time for both, since Sullivan was doing about as well in serious music as was Gilbert in serious theater. The Gilbert and Sullivan series of smash-hit musicals that followed made them household names and earned them noble titles. In the main, the shows satirized the social issues of the day (like women's rights, the military, the aesthetic movement, the law). The pair's second effort, *HMS Pinafore,* was greeted in America (said a reviewer) with "enthusiasm bordering on insanity." *The Mikado* (1885) ran for an unprecedented five thousand performances.

Sullivan also worked with another librettist named Fred Weatherly, one of the most prolific songwriters of all time, with more than three thousand titles to his name, including all-time greats like "Roses of Picardy" and "Danny Boy," as well as less-well-remembered numbers such as "Brown Eyes under the Moon." Weatherly had started out as a criminal lawyer. And one of his most profitable musical partnerships was with a composer who, ironically, may well have gotten away with murder. In 1892 Weatherly and Steven Adams wrote "The Holy City," followed by "Nancy Lee" (which sold one hundred thousand copies). Adams's real

theaters with a fine mist. Leading his surgeon colleagues to exclaim, ambiguously, as they approached the patient: "Let us spray." Lister's trick seemed to do the trick, and by 1879 antisepsis was accepted as a fact of life, not death. About this time, Lister was also experimenting with an idea of German researcher Robert Koch's that heat sterilized instruments better than chemicals did. This, just before Koch made his international reputation by finding the tuberculosis and cholera bacilli and attracting everybody to his classes. This included a young Dutch medic named Christiaan Eijkman, who then sailed for Dutch Indonesia, where in 1890 he would make his own Nobel-earning discovery: that chickens with the staggers were suffering a dietary deficiency. The same went for people, i.e., there were substances a body had to have to stay healthy.

END TRACK ONE

name was Michael Maybrick, and at the height of his musical career in 1889, his brother James died under mysterious circumstances. Arsenic-covered fly papers were found soaking in water, and the house contained a large quantity of the poison. James's young American wife Florence (accused of poisoning James with arsenic) was having an affair. At one point, the prosecution's argument suggested that an adulterer could easily also be a murderer. Nothing was said about the fact that James had had a mistress for thirty years, had fathered several illegitimate children, and was an arsenic addict (at the time, arsenic was used as an aphrodisiac) and that the arsenic in the house belonged to him. At the trial, evidence provided by Michael Maybrick was damning, and Florence was convicted and sentenced to hang. In the face of public outcry, the sentence was later commuted to life and, later still, to fifteen years. The suggestion now is that Maybrick may have framed his sister-in-law in revenge for her affairs and to make sure the Maybrick family would get her inheritance.

The toxicologist acting for the prosecution was Thomas Stevenson, noted forensic expert from Guy's Hospital, London, for whom poisoning was a specialty. The year before the trial, he had taken on a young assistant, new grad Gowland Hopkins, whose exam results had impressed Stevenson. Hopkins later went on to specialize in uric acid, which (as you might expect) led him to work on butterfly wings and the bombardier beetle, before he turned his attention to the effect of diet on the excretion of uric acid in urine. It was during this work that he first noticed how baby rats would lose weight while being regularly fed, so long as their diet did not contain any milk. Further investigation convinced Hopkins that besides fats, carbohydrates, and salt, there must be "unknown" substances that were also essential to health. He called these "accessory food factors" and went public on the matter in 1912.

END TRACK TWO

AND FINALLY . . .

In 1929 Eijkman and Hopkins were awarded a joint Nobel for work which led the way to the discovery of vitamins.

FIVE

1792: JUNIPER HALL TO JET AIRCRAFT

About twenty miles south of London, nestled (as the tourist guidebooks would have it) in the Mickleham Valley which wanders among the wooded chalk hills of Surrey, stands stately Juniper Hall. Originally a seventeenth-century coach inn, it was fashionably remodeled in the mid–eighteenth century and boasts a neoclassic portico and a "sculptured" drawing room complete with posh plasterwork in gold and white. If the place weren't a Biology Field Center, it'd be just the spot for a quiet weekend getaway.

TRACK ONE

The Count of Narbonne got away to Juniper Hall in late September 1792, just in time to avoid a possible loss of his head back in revolutionary France. He'd had a silver-spoon-in-mouth life, had dozens of mistresses and serious debts, was possibly the illegitimate son of Louis XV, a school pal of Louis XVI, the father of Germaine de Staël's son, and a general without ever seeing a battle. In 1791 (just before things went democratic), he was minister for war, for all of three months, during which time about all he did was sanction the printing of the latest infantry training manual. This was just as well for General Winfield Scott, who in 1814 found himself commandant of a training camp near Buffalo, New York, with nothing but the translation to use. It must have worked, judging by what his well-drilled trainees did to the Brits at Chippewa later that year. Scott had a lively career: two thanks-of-a-grateful-country gold medals and promotions (up to lieutenant general), one failed attempt at the presidency (he lost to Franklin Pierce in 1850), much scorched earth in Mexico (1847), various Indian wars, some Canadian border disputes, and above all the professionalization of the American army between 1812 and the Civil War. As part of the latter, in 1823, he recommended the promotion of Sylvanus Thayer for his effective reorganization of West Point.

Thayer had started there as professor of math in 1810 and in short order realized the U.S. Army was out of date in almost every way. From 1815 he spent two years in Europe, watching, taking notes, and buying books on everything military. In 1817, after being named superintendent back at the Point, he boosted the curriculum with science, engineering, and hard work, and cut class size, invited regular inspection by outside experts, and started turning out real soldiers—except in one case, when West Point's loss was Gothic horror's gain. In 1830 cadet Edgar Allan Poe got himself court-martialed and went from bad to verse. Poe then wrote reviews for various journals up and down the East Coast, while regularly falling in love, bingeing, attempting suicide, suffering depressions, turning out poems and tales of supernatural terror, and (with *Murders in the Rue Morgue*) inventing the detective story. When he died of drink on the way to his wedding, Poe left few friends, thanks to his jugular-vein literary reviews, in which the worst insult was to have your writing described as "Channingese." Poe coined the term after reading William Ellery Channing's first book of indifferent poetry in 1843. Poor Channing (patrician, Harvard drop-out, lover of country walks and transcendental discussions with pals Emerson, Thoreau, Alcott, and Hawthorne) spent his life commuting into Boston to visit the library and take notes.

Juniper Hall neighbors Molesworth Phillips and his wife Susanna were frequent visitors, fascinated by the exotic French exiles staying there. Molesworth was a captain in the marines and in 1776 was with Captain Cook on the three-year South Pacific voyage that ended with Cook's murder on a beach in Hawaii, a few yards from Molesworth. Not long after the French had left Juniper Hall, the Phillipses met Charles Lamb, a gentle soul who wrote humorous newspaper articles. Three years later, Lamb was to be commissioned by radical reformer William Godwin to write the work that made him famous: *Lamb's Tales from Shakespeare.* Lamb met all the Romantic hopheads, one of whom, Robert Southey, became a close pal.

Nobody reads Southey any more, except perhaps for his *The Inchcape Rock.* "Varied" is the best word for Southey's output, which included *History of Brazil, Essays Moral and Political,* orientalizing poems, and a translation of *El Cid.* Southey was also close pals with Romantic poet and thinker Coleridge (they married sisters and nearly set up a commune on the Susquehanna River). Coleridge took opium and Southey sniffed laughing gas, so they must have been great company. Southey and his wife moved to Coleridge's village, in northern England in the middle of nowhere, and became known as the Lake Poets, in an area where there was little to do but watch the rain and read books.

In 1825 a party of readers turned up at Southey's place, led by a precocious young astronomer, George Airy. Cambridge professor at twenty-five and astronomer royal at thirty-four (for forty-six years), George became the epitome of overworked officialdom. On government commissions for everything from weights and measures to sewers, coinage, transatlantic cable-laying, the Maine and Oregon border disputes, and railway-gauge measure. And that's not the half of it. Airy produced a zillion papers and won an equal number of medals and awards. He left a mountain of paperwork, neatly filed (his obsession). And in 1826 he went on one other reading party, to Orleans, France. A side trip to Paris bumped him into several French scientists, one of whom was C. S. M. Pouillet. Pouillet discovered that some metals lose their magnetism when heated, and he measured the amount of solar energy hitting the Earth's surface (with a thermometer in water in a blackened container). Other-

His namesake uncle did a lot better. The elder Channing was America's greatest Unitarian intellectual, and inspired a generation of ministers at home and in Europe. He was too serious to be fun, perhaps, but these were testing times for people of conscience. So thought one of Channing's Harvard classmates, pal and Unitarian minister, Joseph Tuckerman, who in 1812 was the force behind the Boston Society for the Religious and Moral Improvement of Seamen. Tuckerman was powerfully influenced by reports of urban poverty in Glasgow, Scotland, tried and failed to get something going in urban India, and then succeeded in urban Boston. He spent most of 1833 opening urban ministries in London and Liverpool, and during the trip he made up for his Indian failure by inspiring the well-heeled Brit Mary Carpenter, who was philanthropy looking for an excuse. Mary ended up taking four trips to India, to spread the educative word. Young wanna-be criminals were her speciaity. In 1846 she opened one of the first "ragged schools" (for street children), in 1851 wrote the influential *Reforming Schools,* and in 1852 opened two reform schools and ran the first Conference on Juvenile Delinquency. Many of her recommendations found their way into legislation. Her right hand in much of this was a chap whose middle name was reform. Matthew Hill went from a teaching background via journalism to the law, and when Mary met him was a radical counsel working primarily on civil-rights cases. After a three-year spell in Parliament, he straightened out the parole system and worked for prison reform.

His brother did the same cleanup job on the mail. Rowland Hill investigated the post office and found chaos and rampant corruption involving special-privilege politicians and aristocrats. Postal bureaucracy was top-heavy even for Victorian England. Hill pushed through an entirely new idea: a prepaid, one-price-fits-all adhesive stamp (known as the Penny Black) that licked virtually all the problems at a stroke. This, in spite of entrenched opposition from the harrumph post office pen pushers, one of whom (he and Hill never hit it off) became a writer of a different sort. His first novels sank like stones (ever heard of *The Macdermots of Ballycloran* or *The Kellys and the O'Kellys?),* but once he'd found his subject (life in English cathedral society), Anthony Trollope never looked back. Twenty-five hundred words a day between 5:30 A.M. and breakfast, his stuff isn't everybody's cup of tea. Henry James said the novels displayed "a complete appreciation of the usual." Trollope's books were garnished with illustrations by an already famous painter, John Millais, who wanted to reform what he saw as moribund, chocolate-box British art. Together with a bunch of like-mindeds (Holman Hunt, D. G. Rossetti, et al.), Millais set up the Pre-Raphaelite Brotherhood to do the kind of painting we moderns would call chocolate-box British art. Getting back to

wise Pouillet did little of interest—except make two major career mistakes. One, in 1849, when as director of the Museum of Arts and Crafts in Paris, he was fired for sheltering a bunch of antigovernment rioters in the building. Error two: Back in 1827, he had accepted the job of teaching the two sons of King Louis-Phillippe. This was OK till the revolution of 1848 dumped the king and Pouillet (tainted by association).

Louis-Phillippe had started out well. A former teacher, he'd been elected to power, wore a top hat, and carried an umbrella. Bourgeois to his back teeth, he had execrable taste in the arts and aimed at keeping the middle class happy, which alienated both far right and far left. After problems with a succession of weak but liberal ministers, he found one to his liking. François Guizot's motto was: "Work hard, make money, and save it." As a professional historian, he was an intellectual heavy, with works like *A History of the Origins of Representative Government* on his résumé. None of this helped when, inevitably, the chips were down—nor did his liberal law establishing compulsory primary education. He had, after all, done nothing about laws outlawing strikes or the property-owning requirement for the qualification to vote. So when a recession in 1847 put workers out on the streets, by February 1848 workers were out on the streets. Guizot was forced to resign, as was Louis-Phillippe the next day. The revolution of 1848 was supposed to be the beginning of the end for the old ways. At least, that's how political reporter Friedrich Engels saw it when he had predicted the downfall of Guizot in a newspaper article the year before. Poor Engels never got the proper recognition for introducing Marx to dialectical materialism, the Romantic-movement idea that everything progressed through the conflict of opposites that brought the emergence of the next stage of development. In 1842 Engels was working in a cotton mill in England where (thanks to the Industrial Revolution and unspeakable living conditions of the workers) Engels expected the revolution would happen first, with a helping hand from him and Marx. While Marx snoozed over his *Communist Manifesto* notes in the British Library, Engels got down and dirty in the trenches with the proles, and by spreading the word with articles in the *Northern Star,* the mouthpiece of the Chartists (a movement with crazy beliefs like votes for all, salaried M.P.s, and secret ballots).

pure-nature-as-is, the Brotherhood picked up where the fifteenth-century Italians had left off. Medieval knights, rescued maidens, early-Renaissance-with-everything, these ingredients made Millais a fortune. He became a member of shooting parties in Scotland where movers and shakers met to slaughter things, and in Millais's case, to sketch glen backgrounds for the next in his never-ending series of canvases.

Curiously, one of his earliest portraits was of Sir John Fowler, a man later associated with never-ending painting of a different kind—that of the Forth Bridge. The bridge is so big that when the maintenance crews finish painting, the work has taken them so long that they have to go back to the other end and start the next coat. Fowler got up to (and down to) a wide variety of engineering efforts: many years spent digging the London subway system, a visit to Norway to study its railroads for the Indian government, a lengthy consultancy excursion to Egypt, and last, but far from least, a successful campaign to persuade Garibaldi to give up a crazy scheme to divert the River Tiber away from Rome. His partner in the Forth Bridge work, Benjamin Baker, also dug a lot of the London subway and was a natural for the Forth Bridge job, having written two can't-pick-them-up monographs *(Strength of Beams* and *Long-Span Bridges).* Baker would later also consult on the St. Louis Bridge and the Hudson Tunnel. Meanwhile, he turned toward Egypt—first, designing the cylindrical ship that brought Cleopatra's Needle (an obelisk) to London, and then in 1902 collaborating on the Aswan Dam with William Garstin. For the previous sixteen years, Garstin had been working toward providing the farmers of the Nile Delta with a better irrigation system than the obliterate-everything annual Nile flood. Garstin's canal system would only work, however, if the annual flood were regulated in some way. Hence the need for the Aswan Dam.

One of Garstin's other jobs was running Egyptian antiquities. In 1899 he put those of Upper Egypt in the hands of a young artist who'd spent four months taking excavating lessons from archaeology guru Sir Flinders Petrie and then six years drawing scenes and copying inscriptions from the mortuary temple of the female pharaoh Hatshepsut. His name was Howard Carter, and in 1922 he made Egyptology front-page news all over the world with his discovery of the unparalleled golden treasures of the tomb of Tutankhamen. Back in 1905, Carter had persuaded a rich visiting Brit to put up the funds to dig (and then preserve) tombs in the Thebes necropolis. After three years, the work was finished. Carter's new backer made sure of leaving the tombs safe from robbers by putting them behind posh new metal gates. Metal, after all, was his thing: Robert Ludwig Mond had made his fortune running the Mond Nickel Company—which had been

The *Northern Star's* owner was Feargus O'Connor, a supporter of Irish independence, a poor people's lawyer, a radical, and briefly an M.P. who lobbied for factory reform and an end to flogging. It was O'Connor's open support for violent protest and his intemperate language that inspired William Booth to found the Salvation Army, which frightened most of the liberal-leaning middle class to stay clear. Lack of their support ultimately led to the Chartists' failure, though in the end Chartism went down the toilet because O'Connor couldn't keep books to save his life (or the movement's finances). One early Chartist casualty was a young Scotsman who in 1842 evaded the police by running away to the United States, where he kind of turned his coat. By the 1850s, Allan Pinkerton was running his own private detective agency and (on behalf of the Illinois railroad) infiltrating the same kind of radical workers' groups to which he'd once belonged. When one of his railroad employers became Union Army General George McClellan, Pinkerton became head of a new government agency (using all his own company security people): the Secret Service, whose main job was keeping tabs on Confederate undercover agents during the Civil War. Pinkerton did a good job, most spectacularly in the case of the beautiful Southern-beauty-society-hostess-turned-spy, Rose Greenhow. Greenhow turned the heads of senators and army officers with secrets to reveal, and recruited a bevy of female agents who would turn their hand from seduction to carrying messages in their fancy hairdos. The Union defeat at Bull Run was all thanks to Greenhow's work, and when the General responsible for the debacle was removed in disgrace, his replacement was McClellan. In no time, McClellan had Pinkerton on the case, and by August 23, 1861, Pinkerton had Greenhow in his clutches. Once in jail, however, Greenhow became enough of a celebrity to be an embarrassment, and in 1862 she was released, sailing to Virginia with champagne and sandwiches.

Later that year, Greenhow moved on to Europe to help with the Confederate war effort. In Paris she met Matthew Fontaine Maury, Confederate commissioner to France and England. Maury was busy on a secret mission: to buy a cruiser for the Confederate Navy. In 1861 William Denny, shipbuilder on the Clyde, sold him a ship that had been launched officially as the *Japan* and given a clean export certificate by British authorities. Off the French coast, Maury rendezvoused with the ship, transferred armaments and

founded by Robert's dad, who in 1899 had developed a process for extracting and purifying nickel.

By Robert's time the company specialty was nickel-chromium alloys that would take anything you threw at them. The worst of which would be a destructive combination of heat and electricity. Undaunted, in the late 1930s, the company produced Nimonic 75, which had high electrical resistance and survived temperatures of 750° C. This stuff was great for the wiring in industrial furnaces. So, onward and upward. In the early 1940s came Nimonic 80, which would take 800° C and three tons pressure per square inch. But there was no use for it—not for a couple of years, anyway. After which it would again be onward and upward.

END TRACK ONE

supplies, and as the CNS *Georgia,* she sailed on into the South Atlantic to harass the U.S. merchant fleet. Maury already had an illustrious prewar naval career behind him, based on his work setting up the U.S. Naval Academy and writing *Sailing Directions for Sailors* (1851). The latter came about after he had found tons of information (in ships' logs held in navy archives) on currents and winds encountered by U.S. mariners over many years. Compiling the data (and adding more, gleaned from skippers in return for the promise of a free copy of *Directions),* Maury identified a number of oceanic tracks that took advantage of the winds and currents to cut journey times in some cases by a quarter.

At 10 P.M. on April 14, 1912, while following one of Maury's tracks, the *Titanic* struck an iceberg off Newfoundland and sank with the loss of 1,517 souls. One immediate result was the establishment, by the American and British authorities, of ice patrols. In 1913, the British patrol ship *Scotia* carried a meteorologist, G. I. Taylor, and his weather tool collection of balloons and kites. After a six months' cruise among the floes, the ship returned to Britain and Taylor went back to Cambridge, to invent the modern anchor used by every small boat in the world today—and to work on soap bubbles. Together with a colleague, A. A. Griffith, Taylor made significant discoveries about how to use soap bubbles to measure the shearing when metal is put under torsion. Griffith moved on to investigate how metals behaved under all kinds of stress. In the early 1920s he was studying the metal and highly stressed blades of a gas-turbine compressor. He realized that the blades were working at much less than peak efficiency and that this could be improved if the blades were designed aerodynamically, to function like little wings. So Griffith designed an axial-flow compressor, capable of driving an aircraft propeller.

END TRACK TWO

AND FINALLY . . .

In August 1928 Griffith met Frank Whittle, who had a crazy idea for an engine that would use a turbine—but not to turn a propeller. The kind of conditions the compressor would be required to withstand turned out to demand compressor blades made of Mond's Nimonic 80. When it all finally came together in April 1942, the new axial-flow engine powered the first British jet airplane, the Gloster E29.

SIX

1750: SMALLPOX TO BIG BANG

In the late eighteenth century, chances were that you'd catch smallpox and die. The disease was endemic in India and the Middle East, had been introduced to Europe by the Crusaders in the thirteenth century, and passed on to America three hundred years later. The only defense (practiced by the Turks) was to be inoculated with smallpox pus and hope that it gave you immunity. By 1720 this was also being tried in England. But even after Turkish-style inoculation, chances were that you'd catch smallpox and die.

TRACK ONE

Smallpox was finally eradicated in England (and then elsewhere) by a country doctor, Edward Jenner (he did his rounds on horseback in silver spurs), who noted that local milkmaids got cowpox and then not smallpox. Was cowpox a safer way? In 1796 he gave the son of one of his landless and poverty-stricken laborers cowpox and then smallpox, and (luckily for Jenner) the kid survived. The news spread like smallpox, and by 1802 the British government had awarded Jenner about three million dollars (in modern money) for his efforts. By 1809 there was a National Vaccine Institution, and Jenner had another five million dollars. After which, he retired to concentrate on his real love: cuckoos. The National Vaccine Institution had benefited from the energetic support of one of those people who have lots of friends because they're able to spread around what most people want: jobs and money. George Rose was one such peddler of influence, as will be clear from his résumé: secretary to the Board of Taxes, secretary to the treasury, M.P., friend of the king, treasurer of the navy, trustee of the British Museum, and almost Chancellor of the Exchequer (he turned it down). If Rose couldn't get you a job, nobody could. So when he wanted the National Vaccination Institution to happen, it happened.

The same thing occurred with his second son William, for whom Dad fixed a seat in Parliament just after the boy left school. After four years of politics, young William went into "medieval" poetry writing (he was big pal of historical novelist Sir Walter Scott), after which he hightailed it for Italy, where he started translating Renaissance Italian poet Ariosto's megawork: *Orlando Furioso* (an acquired taste). In 1817, while passing through Venice, William delivered a box of tooth powder to the dissolute Lord Byron. Byron at this time was in need of more than healthy teeth. Having left England forever the year before ("in bad odor" would be the understatement of all time), what Byron required by now was a cure for gonorrhea. The condition was not ameliorated by a nonstop series of affairs (with everybody from countesses to his landlady's wife), indulgence in general debauchery, drugging up, and being (as was said at the time) "mad, bad, and dangerous to know." This clubfooted, Romantic crazy was also writing some of his greatest stuff: *Don Juan* and *Childe Harold.* Another of Byron's frequent English visitors was William Bankes, an old pal from university days, who turned up on the way home from a tour of the Middle East, lugging a twenty-two-foot, six-ton obelisk he'd "bought" up the Nile. Although it would eventually end up in Bankes's Dorsetshire estate in England, the obelisk's inscriptions had already been copied in Egypt, before removal, by a crafty French goldsmith, who then circulated the drawings in Paris.

TRACK TWO

In July 1800 smallpox vaccine arrived in the hands of Benjamin Waterhouse, Harvard's first professor of medicine, who made enemies when he tried to monopolize its distribution. Waterhouse was also unpopular because after returning from seven years of study in London, Edinburgh, and Leyden (where he became pals with Ben Franklin and the Adamses), he out-gunned the local medical yokels and made no attempt to hide the fact. One of his few friends seems to have been a Newport school chum named Gilbert Stuart, who back in 1775 had also lit out for the bright lights of London and become a hotshot portraitist who then never ran out of clients. If only the same could have been said about his money. After causing a sensation with a portrait of one of his sitters skating (a contradiction in terms, back then), he then painted the London cultural elite. Then his debtors caught up, and he left for Ireland. More top people sat. More debtors. Stuart then left for the States, where he painted five presidents, a Supreme Court justice, and anybody else who mattered.

Unsurprisingly Stuart had begun penniless after arriving in London, and in 1777 he appealed for help to Benjamin West, the American dictator of English art. By this time West was a really big artistic cheese: History Painter to King George III, he had helped found the Royal Academy and had blown everybody away when his first major work, about the recent death of General Wolfe at Quebec, caused a storm with subjects in modern dress. West's history painting established a genre that set a new standard, so although he was moving away from the art world's neoclassical love affair with anything Greek and Roman, his advice was still sought when monumentally important historical matters were under discussion. Such as that of the Parthenon, in Athens—the target for the seventh earl of Elgin, who was about to plaster-cast some of it to bring back and use as models for his posh new country house. In the event, as you probably know, he brought back a lot more of the Parthenon than plaster casts. And the whole escapade cost so much it put him and his family in financial hock for more than the rest of Elgin's life. Still, the Elgin Marbles did good things for the public appreciation of Greek culture. In charge of the on-site drawing and measuring was an Italian painter named Lusieri, to whom Elgin

One of the inscriptions was of the word "Cleopatra" in Greek, chiseled on the obelisk base. Just what Jean-François Champollion needed to show him the way to achieve his dream (announced at age eleven) of being the first to read Egyptian hieroglyphics. Champollion peered and peered, and sure enough there was also a hieroglyphic on the obelisk that turned out to be "Cleopatra" in ancient Egyptian. By 1822 Champollion had done enough of this compare-and-deduce work (with other pieces like the Rosetta Stone, which carried an inscription in both Greek and Egyptian) to have become the acknowledged Egyptian code breaker, was about to get the College of Paris chair of archeology specially created for him, and was well into the dictionary and grammar of ancient Egyptian that would take the rest of his cryptological life. It should in fairness, however, be noted that he also had time to have five children. In 1825 Champollion visited an archeological site in Nola, near Naples, where his new patron and provider of funds, antiquity-lover Louis, duc de Blacas, was digging. De Blacas had a career like a cuckoo clock: now in, now out. Starting in 1789, he was a mercenary for the French, then the Russians, then the Austrians. He was a close supporter of Louis XVIII, and was in exile with Louis till 1814, during Napoleon's first reign; then in France till Napoleon's escape in 1815; then in exile again, till Napoleon's defeat at Waterloo, later that year; then back in power again under Louis, and then under new King Charles X. That is, until 1848 and revolution. Then he was in exile for good to Vienna, which was where he fell ill and, in 1866, was diagnosed as "about to die"—and died.

Such diagnostic precision was standard procedure for Joseph Skoda, Czech medical star of percussion and ascultation (tapping and listening), who used his physics to turn this arcane art into diagnostic technique. Sounds were full or empty, clear or muffled, drumlike or not, and high or deep. This worked well enough to make Skoda rich enough to fund his cousin's future car factory. Skoda began his medical career in 1831, the year cholera struck and killed tens of thousands, so he was hot stuff on preventive medicine. As was his pupil, Ignaz Semmelweis, later working in the obstetrics clinic at Vienna in 1849 and who reckoned newborn and new-mother deaths from puerperal fever were caused by infected particles on the hands of the medical students who were examining mother and child immediately after emerging from the dissecting rooms. Washing hands with chlorine reduced the incidence of death tenfold. For some odd reason, Semmelweis wouldn't publish his data, so Skoda did it on his behalf. Semmelweis's contagion theories (later confirmed by Pasteur) corroborated those of an American who'd got there first (and also hadn't published): Oliver Wendell Holmes. Holmes spent time as a medical student in Europe between 1831 and 1836, and after ten years of

had been introduced (while in Palermo, on his way out to Greece) by another great purveyor of classical pieces "bought" or removed from archeological sites: Sir William Hamilton. Hamilton put up with his wife cuckolding him because she was doing it with the one-armed, one-eyed Great British Hero, Admiral Horatio Nelson.

Emma Hamilton was no better than she should have been. She started life as a serving maid, then graduated via hooker to aristrocrat's mistress, and was traded by the aristocrat to his uncle, Sir William. When Nelson met her and they became lovers, she was already running to seed. Hamilton was recalled to England in 1800, and the three of them traveled together (she, already two months' pregnant by Nelson), stopping off at the court in Vienna, where the populace went ape over Nelson (he had licked the French, the Austrians hadn't). Royal junketing was the order of the day and night. That, and several sessions when Emma sang, accompanied by the great Haydn, who is said to have enjoyed it all so much he dashed off a piece in honor of Nelson. Haydn was known as "Papa" by all and sundry, though in real life he was far from being an old fogy. At various times spy, impresario, hotshot businessman, and close pal of the imperial family, by the time Nelson et al. fetched up in Vienna, Haydn had also just ensured his musical immortality with *The Creation* and with what would, one day, become the German national anthem ("Deutschland über alles"). Haydn had also been making a few bucks on the side with arrangements of British folk songs he'd collected from other composers (or written himself) and then sold to Scottish publisher George Thomson. Thomson had persuaded other worthies like Beethoven, Weber, Burns and Sir Walter Scott to chip in with music and words. In this way, Thomson spent sixty years putting together three hundred songs, in six volumes, with the idea of cashing in on the renewed interest on the part of the English in the traditions of Scottish Gaelic culture. A culture which the English had so recently tried to expunge by fire and sword (so as to replace unprofitable people with profitable sheep, and which is why there are so many Scots in Canada).

Thomson also edited the early poems of another author who had the same Scots' revival in mind. Anne Grant's work evocatively (and perhaps a little romantically) described life among her husband's mountain parishioners. This, when she wasn't harking back to her childhood years

practice settled down to a Harvard professorship and a life of literature: three novels, two books of essays, and many poems. From 1857 Holmes contributed columns to the *Atlantic Monthly:* "The Autocrat at the Breakfast Table," "The Professor at the Breakfast Table," and "The Poet . . ."

One evening in 1865, Holmes regaled the Boston Union Club dinner with a poem to its guest of honor, the about-to-be first U.S. admiral, national hero David Farragut. Farragut achieved everlasting fame on the basis of one battle following a long and lackluster naval career. In April 1862 Farragut led his squadron of Union Navy ships up the Mississippi and captured New Orleans. In January 1864 came the coup de théâtre. As they entered the channel to Mobile Bay, under fire from the guns of Fort Morgan, Farragut's watch shouted "Torpedoes!" (mines). To which Farragut replied, in pure Hollywoodese: "Damn the torpedoes. Full steam ahead!" He won the battle and made the record books when Congress invented his new rank in 1866. Farragut's capture of New Orleans turned the area into a war zone and made life difficult for the local Choctaws, which Farragut discovered only when a priest, Père Rouquette (who looked and acted like a Choctaw) appealed to him for medical supplies. The first Native American priest of American Louisiana, the French-educated Rouquette started as a bishop's secretary but soon turned to Choctaw missionary work, in which he spent the rest of his life, living like a Choctaw and writing like a Frenchman—lyric poetry and prose about the ascetic life and Native American culture.

One of these pieces took the fancy of the one-eyed assistant editor of a local paper, *The Item.* Editor Lafcadio Hearn was thirty-two, with a Cincinnati crime reporter's job behind him. In 1881 he moved on to the literary editorship of the *New Orleans Times-Democrat* and a life of cultural commentary. Two years on the island of Martinique and several novels later, in 1889 he read *The Soul of the Far East,* by Percival Lowell, and decided to go to Japan, where he eventually married a local, became the first great multiculturalist, took Japanese citizenship, wrote widely on Japanese life, and never returned to the West. His inspiration, Percival Lowell, came home to fame (he already had a family fortune) as an astronomer, deciding it was his destiny to find life on Mars. His discovery of the Martian "canals" didn't excite other astronomers but galvanized the public imagination. Lowell financed Lowell Observatory at Flagstaff, Arizona, and continued to greater things, predicting the existence of a planet beyond Neptune (it would be Pluto) and funding the search for it, and (of course) predicting life on Mars.

At one point in 1914, staffer Vesto Slipher pointed the spectrograph he was using to analyze the Martian atmosphere at a nebula in

in America, where, thanks to her popular *Memoirs of an American Lady,* by 1819 she had quite a following. One of Grant's admirers turned up that year to pay respects on his way home from a four-year meet-everybody trip around Europe (he even saw the pope), learning all the languages he could as preparation for a professorship at Harvard. George Ticknor was a Boston Brahmin and chum of Madison, Jefferson, and Adams, so there wasn't much opposition when, once back at Harvard in 1823, he introduced a few curriculum reforms (Harvard was pretty stuck-in-the-mud back then) in his brand-new modern-languages department. In 1825 one of the reforms included the extraordinarily unusual idea of hiring a German to teach German. Karl Follen had only been in the United States for a year, on the run from the German authorities because of his alleged involvement in the assassination of a prominent right-winger by members of the extreme nationalist movement. The nationalists had used gymnastics as a front for the military training of young men so that they would be ready to "defend the Fatherland in the struggle ahead." At Harvard, Follen promptly set up physical education classes and built the college's first athletics field. However, by 1828, he was in political vein once again, this time giving speeches and writing inflammatory pieces for the antislavery campaign.

In 1834 Follen got caught up with a visiting English radical, Harriet Martineau, and accompanied her on a tour of the West (as far as Ohio) in the journey of discovery that would lead to her 1837 *Society in America.* During the trip she found plenty to fuel her abolitionist and feminist sentiments. Martineau was a workaholic, hypochondriac journalist dedicated to the Society for the Diffusion of Useful Knowledge. In her case, this meant churning out endless books and pamphlets explaining the good (or bad) of economics, taxation, legal reform, public health, factory legislation, education, and everything else worth agitating about. In the end, she became one of the great and the good—and the overworked. Back home and diagnosed as terminal, in 1844 she went to mesmerist Spencer Timothy Hall, was miraculously cured, and, quitting the opium medication cold turkey, took off for Egypt and camelback. Hall himself did less well. From a shoemaker's son to a self-taught printer, around 1842 Hall had progressed to phrenology (bump-reading). In 1843 he edited the *Phreno-Magnet* maga-

TRACK ONE

the constellation Virgo. One thing a spectrograph does is to split the light from an object into a spectrum. The spectrum that Slipher's instrument showed knocked everybody sideways. The prismatic rainbow of light from the nebula was predominantly red and becoming more so. Red is the color of an object's light when it's retreating, so this result indicated that the nebula was retreating from Earth—at nearly 1,000 kilometers per second! By 1925 Slipher had found many more retreating galaxies than approaching ones. What did this fugitive behavior mean?

END TRACK ONE

zine. In 1844 he was drawing the crowds at phreno-hypnotism lectures in Scotland. From this point, it was a steady decline, both in health and finance, to a poverty-stricken death. But he'd come close. One lecture in Edinburgh had been attended by no less than science megastar Justus von Liebig, who was in Britain at the invitation of his many former students.

Liebig's new chemistry lab at Giessen, Germany (which he opened in 1824 and ran for twenty-eight years), had introduced such fundamental modern teaching and analytical techniques that today it's a museum to his memory. If you like labs, you'll love this one. There's almost too much to say about Liebig's chemical genius, save that he invented artificial fertilizer (thus kicking off agricultural chemistry and solving the problem of nutrition for the new industrial masses), and late in life, while a professsor in Munich, in 1856 he discovered how to put a silver deposit on polished glass.

This may seem of secondary importance compared with feeding the millions, but it was stellar stuff to August Steinheil, German astronomer, who then went into reflector-telescope-making. This, after having invented his photometer: take a small telescope with two movable lenses set on sliders; find a star, using one lens; then find another star with the other lens; slide one lens till both stars are equally bright; how far you slid the lens tells you how much dimmer one star is than the other. Heavenly stuff to Henrietta Leavitt, American skywatcher and star-magnitude freak. In 1912 she found variable-magnitude (pulsating) stars in the nearest nebula (Magellan), whose brightness related to the length of their pulse. This turned out to be the key to measuring how far away they were: a cosmic ninety-four thousand light years.

END TRACK TWO

AND FINALLY . . .

In 1929, at Mt. Wilson observatory, Edwin Hubble (using Leavitt's trick to measure the distance to Slipher's galaxies) made the mind-boggling discovery that the farther away a galaxy was, the redder its light was, and the faster it was retreating from Earth. This became known as the velocity-distance relation, otherwise known as Hubble's Law. What Hubble's discovery revealed was that the universe was (a) vastly vaster than previously thought and (b) getting even vaster. The way was open for the Big Bang.

SEVEN

1784: SANSKRIT TO CYBERNETICS

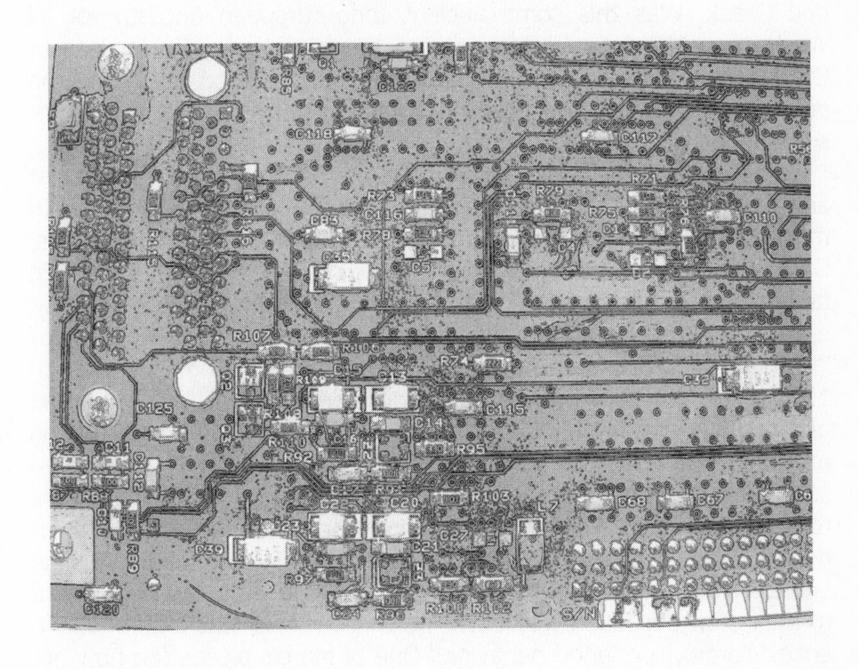

The image of crusty professors moldering in dusty libraries among cobwebbed tomes must have been injected with renewed torpor by the great eighteenth-century discovery of ancient manuscripts written in Sanskrit. This long-not-spoken language of the Indian Brahmins had been on the slide since 500 B.C., and by A.D. 1000 was only understood by a handful of Indian princes priests, and students of ancient epic poems and legal texts. Until the Brits took over India, and in due course decided that the locals ought to be regulated by their own ancient laws.

In 1783, William Jones was given a judgeship in Calcutta. He had already spent several years as an Asian literature scholar with a few publications under his belt, and had turned for a better source of income (what better) to the law. In 1784 he came across a manuscript of ancient Indian legalese, which looked like it'd fit the "indigenous laws" requirement. Jones found a tame scholar to teach him Sanskrit and by 1789 he'd translated *Sakuntala,* a major ancient epic, and was blowing European intellectuals away with the notion that there were strange similarities between Sanskrit and German, Latin, and Greek. Was this some ancient Indo-European ancestor of all Western-speak? Jones's message got heard because he was one of the in-crowd back in London, where he'd been elected in 1773 to the prestigious Literary Club (leading light, the terrifying Samuel Johnson).

Co-newbie was David Garrick, manager of the Drury Lane Theater and the man who turned acting from French posturing bombast to modern scratch-your-ass naturalism. When he met William Jones, Garrick had thirty years of rave-success acting behind him, mostly in farce, and was busy cracking the theatrical whip: getting the audience off the stage during the play, doing away with discounts for latecomers, putting the orchestra in a pit, and installing footlights. In 1776 he (and his wife) acquired a lodger and admirer (and would-be playwright) Hannah More. Who shortly thereafter took a turn for the religious and moved off to a country-cottage life, where she opened schools for the poor to prepare them for servitude and to teach them to read but not to write—point being: writing was potentially revolutionary. And More was no revolutionary. By 1792 she was scribbling endless tracts of the kind that aristocratic employers would pass out to their staff because every other word in the text was either "obedience," "order," or "good behavior." One of More's pieces (on how not to get involved in radical village politics) was grabbed by the government spin doctors, who distributed thousands of free copies. But in spite of this welter of do-good, More became best known (thirty editions in the United States) for a novel.

The hero of this novel was modeled on her good pal John Scandrett Harford, banker and country type, who became a magistrate, county high sheriff, and all those things you do when you prefer things left stuck in the mud. Harford founded a college, drained a bog, and was variously involved in water supplies and cathedral restoration—and was an art buff (went to Rome, bought lots of paintings, and commissioned his own personal copy of the Sistine Chapel ceiling). In 1810 he also ordered the building of an entire village.

TRACK TWO

In 1788 George Forster was librarian at the University of Mainz, Germany, following a checkered career as a naturalist on Captain Cook's world-renowned first world trip and afterward, in London, jumping the gun with a book on the trip ahead of the well-connected (now annoyed) Cook, and then leaving town in a hurry. In search of extra earnings, Forster set up a translation service and, in a fit of do-it-yourself spirt, in 1791 translated into German Jones's English translation of the Sanskrit epic *Sakuntala.* When Friedrich von Schlegel got his hands on the book, it knocked his socks off. At the time, Schlegel was in Jena writing the manifesto for the Romantic movement ("strive for the infinite") at a time (1802) when Romantic Germans were getting excited by the ancient Romantic past. They were also losing every battle to Napoleon, and the snotty French were, of course, taking every opportunity to rub German noses in the we-are-sophisticated-and-you-Krauts-aren't. Schlegel did a quick Sanskrit course in Paris, and in 1808 his *On the Language and Wisdom of India* showed Germans there was an ancient Indo-European culture that might give them a pedigree every bit as posh as that of zee Frogs. Clemens von Brentano (major *literato)* had been on the trail since 1801, collecting old folktales with his pal Achim von Arnim.

In 1803 Brentano moved to Marburg and bumped into a couple of brothers who would raise interest in ancient German matters to world levels (and the personal attention of Hitler and Disney). In 1806 the brothers went collecting for Brentano and then for themselves. What they ended up with, in 1812, became known as Grimm's fairy tales: *Little Red Riding Hood, Sleeping Beauty, Hansel and Gretel, Snow White,* etc. Four volumes of cruelty, incest, necrophilia, xenophobia, torture, and anti-Semitism that (it was discovered by the Grimms) recurred in folktales from all European cultures. By the second edition, most of the politically incorrect material had already been sufficiently sanitized to be suitable, one distant day, even for Disney. However, enough of the love-of-fatherland stuff remained in the narratives eventually also to catch the Nazi imagination. In 1808 Jacob Grimm got the job of librarian for the king of Westphalia, in Kassel. Both of these—the kingdom of Westphalia and the position of king—were as new as the Grimm

Blaise Hamlet (named after his family seat, Blaise Castle, near Bristol) had tasteful stone thatched cottages built for Harford's pensioners and was probably the first real example of public housing in England. The architect who did the work was middle aged and well known for giving the client what the client wanted (Italianate, Tudor, Classical, or—when the client was the Prince Regent and the venue was the Brighton Pavilion—Mughal Indian exteriors with Chinese interior trimmings). The Regent's extravagance gave John Nash the chance to provide London with the tourist venues it has today: Regent's Park, Regent Street, and Regent's Canal. Nash also had a stab at remodeling Buckingham Palace and set the style for ritzy London pied-à-terre town houses for squires-from-the-shires who were up from deepest nowhere for the party season. The best houses were in terraces flanking Regent's Park. If a client's taste was for the medieval and ornamental, Nash turned to his young French draftsman Augustus Pugin, who was hot stuff with Gothic detail, which he would provide for Nash in the form of watercolors done during his field trips with students in Gothic-thick Normandy.

In 1823 Pugin designed the interior of the London Diorama, where you went to marvel at the period's hi-tech equivalent of virtual reality. This involved sitting in an auditorium and staring through large holes in the wall at giant paintings of scenes (mostly urban), rendered amazingly real by the cunning use of lights and smoke. The diorama itself rapidly became one of the sights of new London, and was painted and engraved (for readers who wanted coffee-table glossies to display) in such works as the collection of illustrations of *Metropolitan Improvements* produced in 1831 by Thomas Hosmer Shepherd. Shepherd churned out books of views (street after anonymous street) including one rivetingly titled: *London and Environs.* This included works by the otherwise almost-unknown William Woolnoth, who took "view of . . ." to the extreme, with views of India, China, the Middle East, Scotland, France, and all the cathedrals you might ever hope not to need to visit, now you'd seen the pictures. In 1825 Woolnoth provided ninety-six of these mind-numbingly repetitive pix for *Ancient Castles of England and Wales,* by Edward Brayley, who very soon afterward gave up the delights of antiquity for the fun and games of science. In an age when enthusiasm beat expertise hands down, Brayley wrote and lectured on everything from meteors, to varieties of carbon, geology, zoology, volcanoes, the Ice Age, tides, wind, and the Sun.

At a lecture in 1839, he revealed the amazing new discoveries of one J. Bancroft Reade, cleric, chemist, and magnification freak, who had produced the first teeny-weeny photos (down a microscope) of things like a flea's head. He had done this with paper, silver iodide,

tales were old. Each had been invented by Napoleon the year before, so as to provide a diversion likely to keep his youngest brother Jérôme out of trouble.

The dreadful Jérôme was the runt of the Bonapartes. Appointed rear admiral with the navy in the Caribbean, he tended to keep well out of the way of anything that looked like a fight. Appointed general of the army in Germany, he repeated this avoidance tactic. His personal life was just a little short of indictable. Debauchery, naked theatricals, multiple mistresses, and unnameable sins of the flesh were the order of the day in Kassel. Well before the entire Bonaparte house of cards collapsed with the fall of Napoleon, the burghers of Kassel rose. And Jérôme hightailed it out of there with all the furniture, art, and jewels he could carry—along with his hard-done-by princess wife, who had been hard-done-by from the start, because Jérôme married her under duress after his previous marriage had been annuled by Napoleon. This earlier nuptial was to a Maryland beauty, Betsy Patterson, whom Jérôme had met while in the navy in 1803. The belle of every Baltimore ball, Betsy was Grace Kelly before Grace Kelly. And the marriage contract stipulated that if things went bad, she'd get a third of what Jérôme owned at the time (which, for all she knew, might be France). After Jérôme's big brother cancelled their union, Betsy spent decades in Europe litigating for her alimony—and failing.

The marriage had been conducted by the normally politically adept John Carroll, first Catholic bishop in America (and soon to be archbishop of Baltimore). Carroll had developed his diplomatic savvy on governmental missions trying (though unsuccessfully) to get the Canadians on-side during the War of Independence. He'd then spent years in Italy learning to play the Vatican game well enough for him to return home and in 1805 put American Jesuits back in business (when officially they'd already been given the papal heave-ho). He also succeeded in keeping Roman curial fingers out of the American Catholic pie. In 1806 he commissioned what every bishop, by definition, has to have: his own cathedral.

The man who did the designing, Benjamin Henry Latrobe, knew whereof he built. He'd already completed the Bank of Pennsylvania (first American Greek Revival temple of finance). He'd made various contributions to public edifices in Washington, D.C., including the Capitol, the White

and hyposulfite of soda—just like (and possibly before) Fox Talbot, who later was to get all the credit for developing and fixing. Of course, being Victorian, Reade was familiar with Roman coin molds and fossils found in chalk. Moreover, in 1855 he huckstered the idea that the amazing, recently discovered tree sap, gutta percha (boil it, mold it, let it set hard), was the substitute for everything. Reade's recommendation was to use it as a substitute for glass in a photographic plate. Others had grander plans. In 1856, Curtis Miranda Lampton, Vermont-born London financier, merchant, and socialite, decided to use gutta to insulate the transatlantic telegraph cable he wanted his company to lay under the ocean. When, in 1866, the cable finally made it all the way to Newfoundland, Lampton became a Sir and one of the great and good.

This bumped him into another American great and good, also living in London at the time. Like many humble-beginnings Yanks, George Peabody had made his fortune in trade, advancing from traveling salesman to businessman to import-export maven before moving to London in 1837, where he dealt in canal and railroad company bonds and raised large loans from banks. By 1848 he was back in the United States, helping to finance the war with Mexico. By middle age, Peabody was so rich he was giving it away to build educational institutions and museums, as well as a public housing trust in England. At the height of his financial powers, he managed to persuade investors to weather the panic of 1857 and buy railroad stocks, thus keeping companies like the Illinois Central afloat and its directors (including William Henry Osborn, who'd had investment dealings with Peabody before) from jumping out of the window. The Civil War was Osborn's opportunity of a lifetime, with the Illinois Central's north-south line pointing straight into the heart of Confederate country and ready to take troops, guns, and bullets all the way there. It made him lots of money.

Osborn's son Henry eschewed a railroad career for academia and *two* careers: as a great paleontologist and as a university administrator. He kicked off biology at Columbia and started the Bronx Zoo. He also helped make the American Museum of Natural History a world-class outfit, in spite of his widely publicized white-supremacy belief (see: *Men of the Old Stone Age,* 1915) that humankind had, thanks to miscegenation, been on the skids since Cro-Magnon times, some forty thousand years ago. Henry's stone-and-bones epiphany had happened during a university fossil-collecting trip in 1877. The same trip also inspired a young law undergraduate, W. F. Magie, to switch tracks to science and then get all excited by the 1895 discovery of X-rays.

By 1896 Magie had developed what he called his skiascope

House, and the Navy Yards (and, thanks to the War of 1812, had seen them go up in smoke). Much of Latrobe's work was reworked, but essentially most of what is there today is his. Latrobe had started life in England, and his English cousin Charles Latrobe was one of those footloose types (he would end up governor of an Australian state). In 1824 he went along as tutor to a young Swiss aristocrat, Albert Count of Portales, and together they climbed the Alps, and yodeled, and all that stuff. In 1832 Albert's parents asked Charles if he'd take the boy to the United States (a) to get him away from an unapproved romantic liaison and (b) to have him sow his wild oats where it wouldn't bring disgrace to the family. The two set off and promptly met famous author Washington Irving *(Rip Van Winkle),* who took them around the various sights of New York State. They then headed out across the prairies to the edge of civilization (a.k.a. Oklahoma) for three months of swashbuckling adventures involving irritated Native Americans, wolves, dangerous rapids, sleeping up trees, and the rest. Upon their return to urban comfort and publishers, each man wrote a book (given the hair-raising nature of some of their escapades, Charles's title was the more understated: *The Rambler in North America).* Portales, oats now sown, went home to the life of a diplomat. In 1883 his daughter Marguerite married and became illustrator-in-residence for the publications of Swiss Egyptologist H. E. Naville, who was a caricature of the genre (in the broiling Egyptian sun, he always wore a brown solar topee, Norfolk tweed jacket, baggy pants, pince-nez, and whiskers). Naville never took his nose out of hieroglyphs and tombs. His trench de résistance was the 1893–1896 excavation, on behalf of the Brits, of the mortuary temple of the female pharaoh Hatshepsut.

His young assistant on this big dig was David Hogarth, more of a bookworm than excavator, whose night job was writing travel books with semiautobiographical titles *(A Wandering Scholar in the Levant,* 1896). In 1909 Hogarth was made keeper of the Ashmolean Museum in Oxford and became known for the depth of his knowledge in all things Arabian, having by this time spent half his life in the desert with camel and gun. These esoteric skills paid off in 1915 during World War I, when he looked after various for-your-eyes-only intelligence matters in Cairo, and kept a watching brief on the slightly more explosive activities of his protégé from Oxford days, T. E. Lawrence, otherwise world-famous

("shadow scope"), which allowed you to see, in real time, X-ray views of innards. Two Italians (and Edison) were doing the same thing. Being Edison, he got to market first—with a kit that, in 1898, was just what Harvard physiologist Walter Cannon was looking for to investigate the process of digestion, because with it he could watch a barium meal making its way through a goose's stomach. During this gastronomic event, Cannon noticed that when animals were scared or angry, stomach hunger contractions stopped, only to resume again when the moment of fear had passed. The search for the processes involved led him and his Mexican assistant (Arturo Rosenblueth) to postulate a neural transmitter that would act to return the body's systems to normal after trauma. In 1932 Cannon explained it all for the public in *Wisdom of the Body*, the book that first publicized the concept of homeostasis.

END TRACK ONE

as Lawrence of Arabia. Lawrence succeeded in getting the Arabs to blow up their Turkish overlords but failed to get them to agree among themselves about anything other than that. Apart from the classic Hollywood movie, we know about Lawrence's exploits primarily because of the films shot of him in action by Lowell Thomas, an enterprising American journalist. Thomas's book on their adventures together made Thomas famous, rich, and the host of a long-running (forty-six years) radio news show. Thomas became an American household word, with reports ranging from royal coronations in London to on-board bomber-raid dispatches over Berlin. As one of the new broadcasting megastars, Thomas was on first-name terms with FDR and his science adviser through World War II, Vannevar Bush, engineer and visionary. Bush single-handedly changed the course of American science and technology when he set up the wartime Office of Scientific Research and Development and brought massive governmental funding to labs that would produce innovations in response to wartime requirements. Their work would eventually include the development of the atom bomb. More important, Bush foresaw the postwar need for a national partnership between science and industry that would guarantee American superiority for decades to come.

Back in 1927, at M.I.T., Bush had worked with an irascible mathematician, Norbert Wiener, who during the war, as one of Bush's Office of Scientific Research and Development staff, developed a technique for firing antiaircraft guns more accurately by slaving the guns to incoming radar signals from the target via a statistical equation that would help predict where the target would be in a few moments. This would then point the guns toward the spot where the shell and the target ought to arrive at the same time—with smithereen results.

END TRACK TWO

AND FINALLY . . .

The feedback process involved in Wiener's fire-control system was not unlike that studied by Arturo Rosenblueth during his work with Cannon on the body's homeostatic mechanisms. Wiener and Rosenblueth began to work together, and by war's end, they had come up with a feedback concept that would make possible the greatest scientific revolution since sliced bread. Wiener named the concept "cybernetics."

EIGHT

1610: *SANTA CATHARINA* TO SPECTROSCOPY

In the early seventeenth century, the Dutch boarded the Portuguese ship *Santa Catharina* in the Straits of Malacca and took back to Amsterdam her cargo of spices. Around the same time, the Dutch also started catching walrus in the waters around Greenland and taking their tusks home to make false teeth and knife handles. Not surprisingly the Portuguese and British were less than happy at these incursions. The Brits regarded Greenland seas as their private domain, as did the Portuguese the Indian Ocean.

The Dutch promptly whistled up a young legal-eagle hotshot, Hugo de Groot, and he made their case with his "Open Sea" concept: the sea isn't inhabited, so it can't be anybody's property. De Groot ended up in Paris, where he was appointed Swedish ambassador by Sweden's extraordinary, cross-dressing intellectual queen. Queen Christina abdicated in 1654 and lit out in disguise for Rome, a love affair with Curia biggie Cardinal Azzolino, her patronage of Scarlatti and Corelli, her own personal orchestra and philosophy academy, and the greatest collection of Venetian and other paintings anybody'd ever seen.

In 1710 the art collection was sold to Philip, regent of France, in a deal brokered by Benedetto Luti, the best painter in Rome at the time, who had cornered the art market and was also getting commissions from aristocrats, cardinals, and the pope. That year Luti took on an assistant, a former coach painter from northern England, who was in Rome to learn to copy the great masters. By 1714 William Kent was painting originals and running his own art dealership, selling almost anything to visiting, well-heeled young Brits on their obligatory Grand Tour of Europe, to imbibe vague notions of culture so they'd have something else to talk about back home besides huntin', shootin', and fishin'. One of these hoorays invited Kent back to London, where they lived in the same palace for the rest of Kent's life, and he became tutor to the wife and kids. Once installed in Burlington House, Kent was soon in touch with the vacuous megarich who had nothing to do with their billions but spend them on the stuff he did: neoclassical houses, informal gardens, posh interiors (check out the Kensington Palace grand staircase) and flamboyant (i.e., excessive) furniture. His patron in all this was the trillionaire Earl of Burlington, who was no architectural slouch himself (Chiswick House, London, a small-scale imitation of Palladio's Villa Rotonda, about which some wit noted: "too small to live in, and to large to hang on a watch"). Burlington saw himself as the arbiter of the nation's taste. If it was an artistic pie, Burlington's finger was in it.

That is why he was also director at the Royal Academy of Music, staging performances of the work of German immigrant composer Handel, who, from 1716, also lived in the now-overcrowded Burlington House, and was a favorite with one half of the royal family. This latter matter was to turn out awkward, after Burlington (in Rome again in 1719 on a second G.T.) enthusiastically invited Giovanni Bononcini, by this time a Europe-wide star opera composer, to come to London. In London the Academy commissioned him, Handel, and a third composer each to do one act of a three-act piece. Sides were

TRACK TWO

By 1618 James I of England, losing fish to predatory Dutch offshore herring boats, ordered top lawyer John Selden to rebut De Groot's "Open Seas" argument. Selden was busy, so it took him until 1636 before "Closed Seas" laid the groundwork for the modern-day concept of offshore limit. Unfortunately, Selden's uncompromising defense of the individual's common rights kept him in permanently bad odor with the king and—when he questioned the right of the clergy to exact taxes—with the church. Somehow he survived to die in his own bed, given absolution by one of his pals, Archbishop of Armagh and Primate of All Ireland, James Ussher.

Ussher was one of history's great book collectors, the first "Professor of Theological Controversies" at the newly established Trinity College, Dublin, and (like Selden) loved and hated by everybody: too egalitarian for the established church and too royalist for the Puritans. He also looked both ways when it came to arguing for (or against) the Parliamentary execution of Charles I and the setting up of a republic. He did such good work on church reform that it's a pity that all he's remembered for is his theological chronology (Creation happened on Sunday, October 23, 4004 B.C.; Adam and Eve were driven from Paradise the following November 10; the Ark landed on Mt. Ararat on Wednesday, May 5, 1491 B.C.). After Charles's beheading things got dicey, and Protestant Ussher was offered sanctuary and a pension by Cardinal Richelieu, virtual ruler of Catholic France.

At the time, Richelieu had pretty much completed his great work as Louis XIII's right-hand man. When he'd taken over, back in 1624, the place was going to hell in a handbasket. The economy was down the drain, the nobility spent more time plotting than managing their estates, the royal family was at each other's throats, and the pope was calling all French shots. Richelieu saw his duty as (a) king up, (b) pope down, and (c) any other business. He also taxed the church, founded the Académie Française, and wrote a few bad plays—and encouraged an incoming brain drain. One of which was a crazy Czech Moravian bishop named Komensky, who'd thought up pansophy (a way of harmonizing all aspects of knowledge through universal books), universal schools, a universal college, and a universal

soon taken. Frederick, Prince of Wales, rooted for Bononcini because his father, George II, and sister (both of whom he hated) loved Handel. The rivalry rapidly became the "War of Tweedledum and Tweedledee," and when George II himself got into the act, things really escalated. With all the young aristocrats on Fred's side, Bononcini was getting packed houses, while Handel was playing to little more than royal George. Finding himself well out of his league, Bononcini slipped away to less dangerous climes in 1733, two years after Fred died from being hit by a tennis ball.

In his last years Prince Fred had also taken up with the Earl of Bute, a personable type who shared an interest which Fred (and, more important, Fred's wife Princess Augusta) had in things botanical. Excluded from George II's court, the couple set up in a new house (William Kent, again) in Kew, outside London. Here, after Fred's sporting demise, Bute and the princess started putting together what would one day become the Royal Botanic Gardens. This, logically enough, at a time when explorers and merchants were coming back from exotic spots like America and India with plants nobody'd seen before. All of which needed cataloging and a place to flower. Under Bute's direction, Kew also filled up with temples, pagodas, mosques, Gothic cathedrals, Alhambras, and architectural esoterica of all shapes. When Augusta's son became George III, Bute was in like Flynn and in 1762 (briefly) became Prime Minister—and, as one with George's ear, made enemies.

One of these enemies was another earl who was also a general and who opposed all the decisions being taken that would end up triggering the American War of Independence. Being a good and obedient general, however, when sent, Cornwallis went, and he was better than most of his military bosses. Pity Cornwallis would go down in history as the great loser, since he and his eight thousand men were only stuck at Yorktown in October 1781 because he had been told to stay there and wait for reinforcements. What arrived instead was an army of sixteen thousand assorted French and American troops, plus a large French fleet out in the bay cutting off help and lobbing shells onto Cornwallis and his men. On October 19, outgunned and outnumbered (and with no reinforcements on the way), he made the sensible decision and surrendered. Since events had been brought to this sorry pass by French money and weapons and ships and men and *le général* skulduggery *français* (and everybody knew it), Cornwallis logically surrendered his sword to the French commander, Count Rochambeau, who, ever the gent, passed it on to Washington.

Rochambeau was a veteran of a dozen European campaigns by the time he was sent to head the French expeditionary force in America, and it was his idea to surround Cornwallis instead of attacking the

language. Komensky (who turned Richelieu's offer down) tried three times to set up a college of pansophy: once in England (but a civil war), in Sweden (but no support), and in Poland (but a dead patron). Undeterred, Komensky also wrote the definitive book on education *(The Great Didactic),* as well as the first *See Spot Run* book to teach kids *(The Visible World in Pictures),* and the first modern language course (with Czech and Latin side by side on the page). When he died, Komensky was described as the first world citizen by a German who shared his views and who was himself trying to develop a universal language based on numbers.

Gottfried Leibniz frightens most people because he was probably the last true Renaissance know-it-all. Philosopher, historian, inventor, statesman, librarian, and social reformer, he developed the binary system and, far from least (in a manner of speaking), came up with infinitesimal calculus at the same time as Newton (who got the credit). Leibniz also put his money where Komensky's mouth was: he harmonized the universe in 1698 by postulating fundamental particles before fundamental particles (a.k.a. "monads," energetic entities from which everything was made), existing in perfect harmony as the end product of God's plan ("the best of all possible worlds").

This Pollyanna view of life was music to the satirical ears of Voltaire, who in 1759 took Leibniz to the cleaners, caricatured as Dr. Pangloss in *Candide.* Voltaire had spent much of his life poking holes in other people's obsessions. And since the people in question were often folks such as the regent of France, important ministers of the crown, the French intellectual establishment, and the no-jokes-we're-Calvinist government of Geneva, Voltaire spent much of his time in prison or well out of the spotlight. This tended to be either at Cirey at the château of his mistress, the lovely Émilie du Châtelet (where the couple acted like an intellectual mousetrap and soon had the pointy-heads of Europe beating a path to their door), or Ferney (close to the border with Switzerland, in case of yet another French arrest warrant). It was while Voltaire was at Cirey with Émilie that he heard from an Italian researcher with an interesting experiment. Lazzaro Spallanzani was busy chopping up salamanders, toads, frogs, and worms, trying to work out why they regenerated the chopped-off bits. What may have appealed particularly to Voltaire was the ticklish matter of worm souls. If you cut a worm in half, and the two halves turned

Brits in their New York headquarters. After the dust settled, Rochambeau returned to France, promotion, and glory at the hands of the king. It was not the wisest move, as it happened. Ten years later, Rochambeau would find himself in a revolutionary prison cell, accused of being an aristocrat, and would nearly lose his head. His cell mate would do less well: General Alexandre de Beauharnais—believed to have served with Rochambeau in America—in spite of having been a revolutionary's revolutionary (renounced all noble privileges, was politically radical, ordered the king's arrest), was in the pokey for dereliction of duty after losing the city of Metz to the Austrians. It didn't help his case that he also had a virtual, in-name-only marriage with Marie-Josephe-Rose Tascher de la Pagerie, who came from a quiet little village named Noisy and would become better known as Josephine—the name she used when she became Napoleon's first wife. The unreal marriage between Alexandre and Josephine produced a real son and heir, Eugène, who did a great deal better than his dad.

Apart from the obligatory apprenticeship to a carpenter (a revolutionary requirement for young aristocrats), Eugène's new stepfather (being who he was) made things a little easier. Eugène joined in the coup that made Napoleon first consul, had a good war, and then saved Napoleon from a murder plot, so he was permanently flavor of the month. Napoleon made him a prince of the empire and gave him Italy and the hand in marriage of Princess Augusta of Bavaria—which was where Eugène made sure to be, in 1814, at the time of Napoleon's abdication. Even when Napoleon escaped from Elba in 1815 and made the "hundred days" temporary comeback, Eugène kept his head down (apart from writing letters of support) until Waterloo was over and Napoleon was finally tucked away on St. Helena.

Eugène's father-in-law Maximilian had started royal life as elector of Bavaria, and in 1801, with an eye to the future, had signed a treaty with Napoleon to discourage the predatory Austrians next door. Napoleon had some of the best surveyors in Europe at his disposal and had called together a commission to survey all his conquered territories. He offered their topographical expertise to help Max with the reforms Max had in mind—which needed funding (which would be derived from property taxes) and which required better maps (which would tell him who owned—and therefore who owed—what). Max set up a land registry, and in 1805 turned an expropriated Benedictine monastery at Benediktbauern into a top-secret center for the production of the lenses needed for surveying instruments such as theodolites, telescopes, and the like. By 1810 this place was producing the best lenses on the continent, thanks to its young director of lens-

into two wholes, where, asked Spallanzani, did the soul of the extra worm come from? Trouble with the Vatican was avoided with the suggestion (later to turn into reproductive physiology) that animals contained the seeds of their future children. This didn't quite answer the theological conundrum Spallanzani had generated, but it was good enough. Spallanzani did a zillion other scientific investigations, too, of course (as they all did, at a time when almost everything was yet to be discovered).

Spallanzani became so famous that he was the role model for the science wizard in one of the *Tales of Hoffmann,* by Ernst Theodor Amadeus Hoffmann, a German legal counsel with a theatrical bent and a drinking habit, whose creepy stories of automata, the supernatural, and the undead inspired Delibes *(Coppelia),* Offenbach *(Tales of Hoffmann),* and Tchaikovsky *(The Nutcracker).* Hoffmann wasn't born with Amadeus as a middle name, but Wilhelm, the name change being in honor of his hero Wolfgang Amadeus Mozart. Mozart had also dabbled in the mystical (using one of his friend Mesmer's magnets to treat a character in *Così fan tutte).* Like Hoffmann, Mozart had also done his fair share of story-snatching with *The Marriage of Figaro,* lifted from the play of the same name by a French theatrical failure named Caron de Beaumarchais.

Beaumarchais was a success in other, more important ways (as in: funding the American War of Independence). Just before the tea hit the water in Boston, Beaumarchais had spent time in London on a spying mission, from which he reported back that the Brits were secretly desperate to get rid of the thirteen colonies and all that was needed was a slight shove from France. The shove (millions in laundered money, a fleet of ships, thousands of soldiers, and more armament than the Americans needed) ended up bankrupting France and causing widespread chaos and anarchy known as the French Revolution. But in the meantime, it allowed the French to flip the bird to the Brits—which, from the French point of view, was the original purpose of the entire exercise. American independence came second.

One of the other things that took Beaumarchais's fancy was technology, which was why (as watchmaker to the royals) he was at Annonay in September 1783 with the king and Marie Antoinette to watch the latest Montgolfier hot-air balloon take its first crewed flight. On board: a sheep, a

TRACK ONE

crafting, Joseph von Fraunhofer. Fraunhofer's obsession was to make lenses that wouldn't distort. So he was checking glass for purity when he found that the best test was to use a very fine line of bright, monochromatic light and see if the line was changed in any way by passing through the glass. He produced this bright beam by shining light from a flame through a slit into a telescope. It was when he decided to do the same trick with light from the sun (and passed it through a prism) that he noticed he wasn't getting a simple bright light but that the spectrum was interrupted by fine dark lines: Frauenhofer had discovered Frauenhofer lines.

END TRACK ONE

rooster, and a duck. By 1806 manned balloon flights were ho-hum. That year two French scientists, Gay-Lussac and Biot, went up to twelve thousand feet to see if height did anything to magnetized needles. It didn't. Biot was one of those guys who'd take a flier at anything that caught his fancy. In his case, over a number of years, this consisted of work on mirages, gas density, astronomy, sound transmission, geodesy, fish swim bladders, metal conductivity, and electrolysis. Above all, Biot's fascination was with crystals and what they did to light, which was to rotate its polarity. This meant you could tell things about crystals dissolved in liquid by shining polarized light through the liquid, rotating the light beam, and watching what happened. This meant you could tell how sweet your beverage was before or even without tasting it. Biot's saccharimeter was the sugar industry's absolute cup of tea.

No surprise that Biot would encourage others with the same nerdy tendencies, such as one Alfred-Louis-Olivier Legrand Des Cloizeaux. Nerdy? The rare mineral descloizite is named after him. Biot arranged for Des Cloizeaux to take a fun trip to Iceland in 1845 to study crystals. Des Cloizeaux returned the following year, with a bunch of Germans keen to look at anything that spouted. One of those peering riskily (because he was one-eyed, thanks to an earlier lab explosion) into the geysers and volcanic fumaroles was none other than hot-gas maven, Robert Bunsen. Bunsen was the first to show that English iron furnaces were losing 80 percent of their heat and valuable by-products (like ammonia) up chimneys, and that gas recycling would be a good idea. Bunsen also invented the school-lab burner all school children know and love. The great thing about the Bunsen-burner flame is its nonluminous nature. If you burn anything in it, the flame turns a different color, depending on the burning material.

END TRACK TWO

AND FINALLY . . .

In 1859 Bunsen's sidekick Gustav Kirchhoff shone a bright light through such a burn-event and a prism and explained Fraunhofer's dark lines. He found that the burning material absorbed the wavelengths of light it usually gave off, leaving dark lines where that wavelength light used to be in its spectrum. So now you could find out what something was made of by burning it—unless, like the sun or stars zillions of light years away, it was already burning and saved you the bother. This neat trick became known as spectroscopy.

NINE

1686: POLITICAL JINGLE TO NYLON

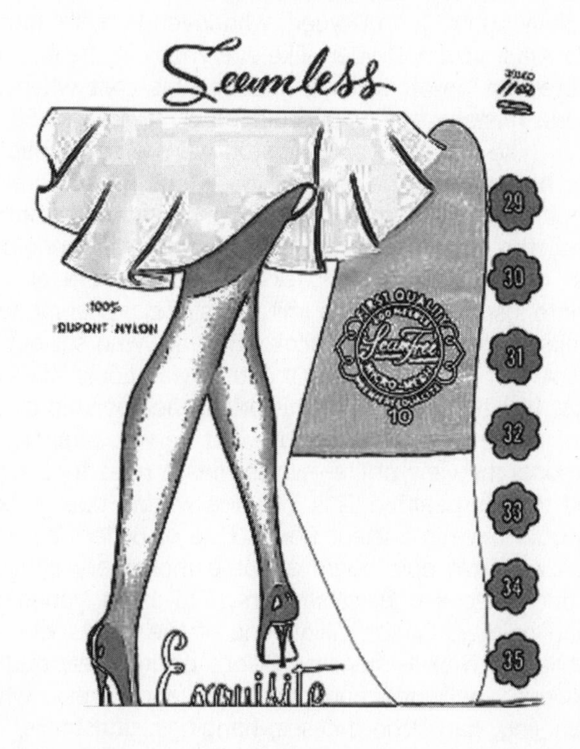

You could not be blamed for listening to the BBC Radio World Service just for the pleasure of hearing that old English air they often play as a signature tune: "Lillibulero." If you don't know the piece, find the wavelength, and you'll soon be whistling away. Interestingly enough, this jolly ditty began life in the seventeenth century as the kind of anti-Catholic propaganda jingle you might hear today in certain quarters of Belfast.

TRACK ONE

Whether or not "Lillibulero" was already a traditional Irish tune, it came to smash-hit prominence as the 1686 arrangement by Henry Purcell, who composed for the royal court. This was like having an open-ended contract with the biggest recording company around. Purcell served three royal bosses: Charles II (anthems, religious airs), James II (coronation anthem), and William III (wife's pregnancy, funeral). With a for-life job as the organist of Westminster Abbey, Purcell was able to turn out theatrical inserts, incidental and major pieces, and all-sung dramas like *Dido and Aeneas,* without having to worry about box-office returns. That was more than could be said for one of his playwrights, John Dryden, who loved Purcell's music for his 1690 drama *Amphytrion.* Dryden, like everybody in the age of sudden Catholic-Protestant power swings, changed his coat when prudence required. One mistake and you could be on the rack. So Dryden's major pieces (like *Absalom and Achitophel)* are so cryptic that you could never be sure who he was talking about. This has also meant that without a background briefing on Dryden's contemporary shenanigans, the modern literature student of his work can suffer drowsiness—not least because back then, they were all squeezing their plots into insomnia-inducing imitations of classical models.

All thanks to the influential Nicolas Boileau, who set out the rules in his 1668 *Art of Poetry.* Boileau's patron was Louis XIV's mistress, who also got him the job of royal historian, membership of the Academy, lots of money, and a big house. So he was supportive of the absolutist-monarchy view of life—which meant rules for everything, if you wanted to stay healthy. This included writing poetry exactly the way they would have in ancient Rome. One of Boileau's interminable classics was a heroic epic poem about a monastery squabble over where to put a lectern. Illustrated in 1713 by a young theatrical designer and painter, Claude Gillot, one of the artists who gave the Louis XV style much of its frivolity. Gillot's designs featured architecture, scrollwork, shellwork, and people in fancy dress, which were used on paneling, harpsichord cases, hangings, gunstocks, etc. Gillot also did much to kick off a new trend known as *fêtes galantes,* portrayals of the new aristocratic way to have fun: dress up as peasants or theater characters and stand around in a park. It was Gillot's pupil Antoine Watteau who, around 1708, raised the technique to a genre with his charming and subtle characterizations of the figures and the very slightly naughty addition of allegorical images to suggest what might be going on, later, behind the bushes. It turns out Watteau probably made money on the side with much more explicitly erotic stuff, commissioned by private patrons. One of these patrons was

"Lillibulero" got words in 1687, thanks to Thomas Wharton, who turned it into a satirical anti-Catholic number that was soon on all lips and helped get rid of the Catholic King James II in favor of the Protestant William III. Wharton was a Protestant and was soon doing well under the new king. He was also an unscrupulous and consummate liar, obsessed with horse racing, and infamous as the greatest rake in the country, excessively concerned with "vice and politics." None of which, in 1706, prevented his promotion to commissioner for the Union of England and Scotland, or (three years later) becoming lord lieutenant of Ireland. On leaving for that job, he took with him composer Thomas Clayton to organize the entertainments in Dublin Castle. The year before, Clayton's *Arsinoe,* the first all-sung opera in the new Italian style, had appeared at Drury Lane. From then on, for his briefly successful career, Clayton disregarded the fashion to do Italian works in Italian, and after he got back from Ireland in 1710, he wrote many settings of English poetry to music.

This effort had the support of the urbane publisher and essayist Richard Steele, who by this time was producing the *Tatler,* a thrice-weekly, one-page review of gossip, politics, literary criticism, and foreign news. Steele wrote much of the material himself. In 1711 the *Tatler* folded, and Steele followed up with the wildly successful (ten thousand copies in one week) *Spectator,* whose somewhat more barbed political satire made enemies that soon brought the magazine down. Steele spent much of his time with the wits of London, in particular at the literature-and-politics Kit-Kat Club. Between 1700 and 1720, the portraits of its forty-eight members, including Steele, were done by Godfrey Kneller, a naturalized German-Brit and one of the most successful portrait painters of all time. Starting when he arrived in England in 1677, for forty-six years Kneller painted ten reigning European monarchs, fourteen admirals, innumerable members of the nobility (and wives and mistresses), and almost anybody important who could afford his fees. And he was made a baronet. Settled in comfortable, if crippled, old age in his palatial country house (now known as Kneller Hall, at Twickenham, outside London), in 1721 he caught a fever and eventually died, in spite of everything his doctor could do—which was surprisingly

very likely the banker and art collector (Rubens, Van Dyck, Titian) Pierre Crozat, who invited Watteau to work in his Paris mansion throughout 1711. Crozat owed much of his success to his rags-to-riches elder brother Antoine, son of a peasant who was good at figures and ended up lending the government loads of money.

In 1712 Antoine Crozat bought the rights to everything coming out of French Louisiana (the territory now covered by Arkansas, Illinois, Iowa, Louisiana, Minnesota, Mississippi, Missouri, and Wisconsin). Five years later, the fabled Louisiana mountains of gold and jewels were still a fable, so Antoine passed the monopoly to John Law, a devious Scotsman who could have sold refrigerators to Eskimos. Law proceeded to pull a similar trick on numbers of greedy French investors who didn't yet know the Louisiana scam was just that. Law had already made a fortune from gambling and had a scheme for saving the French economy from the toilet. Part I (1717): switch to paper money backed by gold, which would get business moving again, and let Law take over Louisiana, all French foreign trade, and the Mint (in return for which Law would pay off the national debt). Part II (1718): pay for all this by selling shares in the Louisiana project. By 1720 investors were making 1,000 percent, and you just know it was about to go bad. Prices soared, inflation followed, then devaluation and collapse. Law legged it over the border. At the height of his fame (controller-general of French finance), he had had a great falling out with his ex-pal, British ambassador in Paris John Dalrymple, who was then recalled (to placate Law). Dalrymple's prime job in Paris had been to keep an eye on the activities of potential troublemaker Prince James Stuart, who back in 1715 had made a failed dash to grab the English throne. The Paris post was a reward for Dalrymple's distinguished military career, and after the Law debacle, he went back to soldiering and ended up as a field marshal.

In 1742, when he was commander of British forces on the continent, Dalrymple appointed an Edinburgh University professor of pneumatics to be physician to his armies. Six years later, John Pringle returned to London to write *Diseases of the Army,* and to achieve fame as the man who first suggested that field hospitals ought to be neutral, thus anticipating the Red Cross. Pringle mixed with the science greats (Franklin, Priestley) and by 1766 was doctor to the king and queen. In 1772, not surprisingly, he became president of the Royal Society, and for six years in the job failed to notice that everything in the Society's garden was not rosy. When his successor Sir Joseph Banks took over in 1778, the way he cleaned up the Society's administration and then packed the membership with his cronies led to mass resignations but left him dictator of British science for the next forty-two years. The formidable Banks had a pedigree you didn't

little, given Dr. Richard Meade's considerable experience. Meade knew Kneller, collected art (Raphael, Michelangelo, Holbein, Rembrandt, and many others), and achieved fame and fortune by treating such luminaries as Newton, Pope, Walpole, and George II.

Meade received all this, after his renowned experiment, carried out as a demo for the Prince of Wales, which involved infecting seven convicts with smallpox and then giving them the new and unproven inoculation recently arrived from Turkey. The convicts recovered and remained immune. So the prince's children took the needle, and then so did everybody else's. Meade also wrote the first major public-health piece on preventing the spread of plague. His single mistake seems to have been to support William Lauder, a splenetic Scots classics scholar and failure, who made waves in 1747 with the announcement that he'd analyzed Milton's *Paradise Lost* and found it to be almost entirely plagiarized from ninety-seven other writers' work. Milton, Lauder said, was a forger. In the uproar that followed, a cleric with a literary bent, John Douglas, spent several weeks in the Bodleian Library at Oxford, chasing up all Lauder's claimed sources and systematically revealed that Lauder himself had forged them. Lauder collapsed, confessed, and left for Barbados. Douglas, reputation made, went on to become a trustee of the British Museum and eventually bishop of Salisbury. Douglas took the odd sally into political commentary and, in 1752, wrote a piece (as cautiously stated as walking on eggshells) in which he showed that modern miracles, like that of "cure by royal touch," didn't satisfy the criteria laid out in the Gospels and so weren't miraculous.

The argument took the form of a letter to an anonymous correspondent, which later turned out to be a guy he'd been at university with: no less than division-of-labor, *Wealth of Nations* economic whiz Adam Smith. During a visit to Paris in 1766, Smith fell in with Abbe Morellet, one of those eighteenth-century French scientific clerics-in-name-only, who knew the entire Enlightenment crowd from d'Alembert to Voltaire, had contributed to Diderot's great *Encyclopédie,* and, being a liberal thinker, was of course pro-American. This, especially after meeting Jefferson in 1785 and working with him on the French translation of Jefferson's *Notes on the State of Virginia.* Alas, the partnership foundered, as literary collaborations often do. Jefferson's consolation was

question. He'd been with Captain Cook on the great voyage, he'd spent some of his vast fortune on building an unequalled botanical collection, and he'd persuaded his pal the king to start Kew Royal Botanic Gardens. In 1779 Banks came up with the genial idea of transporting convicted criminals (sample qualifying crime: stealing a handkerchief) to Australia. In 1786 he helped draw up the plans for how to establish and administer such a colony, and once it was up and running, in 1798 he persuaded a young botanist, George Cayley, to go out to New South Wales and collect for Kew.

Cayley arrived in 1800 and stayed for ten years of botany (all you wanted to know about eucalyptus) and exploration. Including an attempt to find a way across the Blue Mountains, which blocked the colony's expansion to the west. In 1804 he and a group of minimum-security convicts failed to find the way. A flavor of their difficulty can be gleaned from the name they gave one place: the Devil's Wilderness. After more than a week of up-ravines-and-down-again, Cayley named Mount Banks and turned back. In 1813 George Blaxland (also persuaded out to Australia by Banks) used the range ridgelines to find an easier way and crossed the mountains to a point where he saw "enough grassland to support livestock for 30 years." His brother John followed him out to Australia and managed to stir things up with demands for greater freedom of trade and more liberal laws than were usual in penal colonies. He got what he wanted and settled down on an estate outside Sydney. There, in June 1834, he had a visit from a world-traveling Austrian baron, Charles von Hugel, halfway through his six-year tour of the Middle East, India, Southeast Asia, and all points between. High-born von Hugel was yet another plant collector (he would establish the Vienna Horticultural Society), who began and ended his adult life in the army, involved in cloak-and-dagger missions, eventually writing up his extraordinarily varied experiences in a large number of publications.

What he never wrote about was why he'd left home in the first place. Back in 1831, he was engaged to the beautiful Countess Melanie Ferraris when megastatesman Prince Metternich literally snatched her to be his third wife. The second wife had given him a son, Richard, who married another stunner, Pauline. When Richard arrived as Austrian ambassador in Paris, Pauline made instant friends with the Empress Eugenie, and the pair rocked the city with their fun and games—and clothes, provided (shockingly) by a man, Charles Worth, who flounced up extravaganzas in tulle and brocade. Worth then raised Paris eyebrows when he gave Eugenie a walking skirt with a hemline four inches above the ground, which almost revealed the royal ankle.

It was a young designer the House of Worth hired in 1901 who

an ongoing affair with a stunningly beautiful (married) English painter, Maria Cosway, whose husband did a profitable line in pornographic miniatures set into snuffbox tops. Maria found real fame as a promoter of women's education and became an Austrian baroness. Maria's brother, architect George Hadfield, met Jefferson in 1789 and unsurprisingly five years later was offered work on the Capitol in Washington, D.C. Hadfield also became the first naturalized American citizen. With Jefferson's help, Hadfield did well, with various public commissions including Washington City Hall, the Washington Theater, and (his best-known) Arlington House. This last was an 1802 job from George Washington Custis, the great man's adopted grandson—who rarely let you forget it throughout his brief career as playwright of dramas with a national theme. Opening nights would "happen" to occur on his grandfather's birthday. This didn't save Custis's indifferent material from empty seats and short runs. So he retired to his estate and Arlington House. Custis lived in one wing, and the grandfather's memorabilia filled the rest.

In 1864 the estate was bought by the government and turned into what it is today: Arlington National Cemetery. This was the idea of General Montgomery Meigs. After the military academy and a spell as an engineer, Meigs was made quartermaster general of the Union Army, and during the Civil War, he streamlined the entire military bureaucracy and spent the incredible (for then) sum of $1.5 billion, for which he accounted to the last cent. This, in an age of kickbacks, was little short of miraculous. One of the most complex tasks Meigs had to face was to make sure that supplies arrived at the various front lines when they were supposed to. So the condition and operation of the railroads and their associated dot-dash message systems was critically important. For this task, Meigs hired Thomas Scott, former superintendent of the Pennsylvania Railroad, and when Scott arrived in Washington, he brought with him an impressively talented young man whose job was to look after army telegraphic communications. After the war, Andrew Carnegie stuck to railroads, and then iron, steel, and coal. By 1889 he was lecturing on how to become (and remain) rich and then how to give it away before you died (philanthropy was Carnegie's proof of real success). In 1885 one of Carnegie's many bequests went to impresario and benefactor Jeannette Thurber, who'd just started the new

TRACK ONE

went all the way. Paul Poiret did away with the corset, introduced the hobble skirt with split, and (by 1925, after World War I put women in work) raised the hemline halfway up the calf. There was only one way to go from there, and World War II made sure. Cloth rationing pushed the hem to the knee. This left one major problem. Every silk-worm available was making material for parachutes. So there were no more silk stockings.

END TRACK ONE

TRACK TWO

National Conservatory of Music in New York. With the aim of generating a national American style, Thurber offered Antonin Dvorak twenty-five times his normal salary in Prague to come and do for the American sense of identity what he had already done for the Czechs'. Dvorak's *New World Symphony* was the result, and in its use of Native American rhythms, black spirituals, and colonial folk tunes, it set the style for American composers from then on. Back home, at its 1896 Vienna premiere, Dvorak sat next to the man who had given him his start in musical life, the great Brahms. Before he became great, Brahms used to tinkle the ivories with a chamber-music group that included an eminent professor of surgery at Gottingen named Ferdinand Lohmeyer.

In 1892 Lohmeyer's daughter Emma married a member of the Gottingen physics faculty, Walter Nernst, who won the Nobel for his Third Law of Thermodynamics and spearheaded the union of physics and chemistry. In 1900 a young American, Gilbert Lewis, spent a post-grad year working with Nernst. Inspired by his use of physics to explain chemical reactions and the recent (1897) discovery of the electron, Lewis went back to Harvard and took physical chemistry a step forward with the idea that atoms might bond by sharing their electrons. In 1923 his *Valence and the Structure of Atoms and Molecules* laid the groundwork for modern chemistry and excited another American chemist, Wallace Carothers, to publish a piece that caught the attention of the DuPont Company. DuPont offered him a job in research that was more fun than teaching, so he joined the company in 1928. By 1929 Carothers had discovered that chains of molecules known as polymers were held together by simple bonds and that there was no limit to the length you could grow such a chain. While trying to make the longest chain in the world, Carothers produced a polymer that was tough, elastic, and heat resistant—and it could be drawn out into fine filaments and then spun into threads.

END TRACK TWO

AND FINALLY . . .

Carothers's polymer solved the problem of short hemlines and lack of silk. When it went on the market in 1940, his new material was named nylon.

TEN

1703: KIT-KAT CLUB TO SUNGLASSES

The Kit-Kat Club is thought to have been named after the tavern where, in 1703, the first club dinners were held (at the sign of the Cat and Fiddle) in the Temple Bar area of London. The purpose of the club was to meet, greet, wine, dine, and think up ways to embarrass the government. Members were opposition politicians, aristocrats, and the usual literary suspects.

TRACK ONE

Charles Montagu, Earl of Halifax, was a founding Kit-Kat member; a friend of Newton, Locke, and Halley; financial supporter of famous scribblers Congreve and Addison; and the man who set up the modern form of government borrowing from individuals: the National Debt (money borrowed), interest-bearing government bonds (promise to repay), Bank of England (allowed government creditors to borrow against the bonds), and guaranteed taxes (so you were sure the government would cough up). A respite from these weighty national matters (Montagu's mistress for fourteen years) was Newton's niece and in-house hostess, the charming, witty, and beautiful Catherine Barton. As was traditional at the Kit-Kat Club, Montagu used a diamond to inscribe a toast to her beauty on a wineglass. In 1715 Montagu died (leaving Catherine a fortune, which his family then made sure she never got), and Catherine went back to her uncle's place for two years.

That is until a young man turned up, was smitten, and married her. The new husband, John Conduitt, was also so in awe of Newton that he scribbled down every word the great man said to him. After Newton died and Conduitt took over as Master of the Mint (and, in an era of gold and silver coinage, caused a row by suggesting copper money as the answer to the shortage of small change in Ireland), the other thing Conduitt did was plan (but fail) to write Newton's biography. To that end, he collected a mountain of Newtoniana (and his own scribbles), which was eventually used by others who took up the monumental task. Conduitt's job at the mint had, apparently, first been offered to the Reverend Samuel Clarke, a close pal of Newton's, who turned the job down because it was too "secular" for a churchman. This didn't, however, prevent Clarke from accepting a job at the mint for his son. Clarke was considered hot theological stuff and led the argument for a rational form of Christianity at Queen Caroline's twice-weekly egghead chat party. At one of these, around 1724, Clarke met and disagreed violently with George Berkeley, Irish cleric and purveyor of the view that if you don't experience something it's not necessarily there (except in the mind of God). Berkeley spent years trying to raise the cash to set up a college in Bermuda (to take the Gospel across the Atlantic and because the Bermudan climate was pleasant). On the basis of promises from Parliament of (what would today be) a million dollars or so, Berkeley spent three years waiting in Newport, Rhode Island, for the money to arrive, while having an influence on American higher education. In the end no money seemed likely to be on the way (as he himself might have philosophically said: "Out of sight, out of mind"), so in 1732 he sailed home, disillusioned, to a successful career as a bishop.

TRACK TWO

When the Kit-Kat Club was first opened, Lord John Somers was one of its more colorful (and powerful) members. A legal eagle by trade, by 1695 Somers had advanced from solicitor-general to lord chancellor (the nation's top lawyer), had the king's ear, and led the regency council on those occasions when the king was out of the country concluding (unsuccessful) treaties with the French. Only one bright spot adds a little zest to what seems to have been a life of nonstop bureaucracy. In 1701 Somers was implicated in the case of the nefarious Captain William Kidd (whom he had commissioned to chase pirates and who had himself then turned buccaneer, flown the skull-and-crossbones, committed mayhem and murder, and was duly caught and hanged).

Nothing illicit was proved about this common-enough-at-the-time arrangement between one in power and one in crime, but guilt by association was enough for Chancellor Somers to have to resign his chancellorship. Meanwhile, his sister Elizabeth had married another lawyer, Sir Joseph Jekyll, who rose to prominence in government with his Gin Act (a tax on hard liquor), after whose passage through Parliament a bodyguard was needed to save him from the thirsty mob. Jekyll was a liberal who supported moves to outlaw press-ganging (the term for forcible recruitment to the navy, which often involved making prospective volunteers unconscious with either billy club or booze, so that when they regained awareness they were already onboard ship and well out to sea). The press-gang legislation had been first introduced by Jekyll's pal, soldier and member of Parliament James Oglethorpe. Oglethorpe had an even more ambitious plan: to move the poor or oppressed from London to an overseas (American) paradise where, according to Oglethorpe, balmy breezes blew, disease was almost nonexistent, and where the rain would soak trees so much they'd fall down at a touch (thanks to which, clearing land would be easy). And where, since the place was on the same latitude as China, silk could be produced, as well as wine, oil, dyes, and medicines, and who knew what else. This other Eden (which, of course, Oglethorpe had never visited) was to be named "Georgia." Oglethorpe and twenty backers (including Jekyll, who contributed a large sum) were given a charter, and in 1732 off they went, with 120 poor or oppressed English set-

Berkeley left behind the guy he'd taken out with him to be director of the Bermuda College Painting Academy: Italian-trained, unsuccessful portrait painter John Smibert. Smibert found himself a big artistic fish in the small Bostonian pond to which he then moved, and where he was soon overwhelmed with commissions to paint Boston's mercantile elite. Encouraged, Smibert branched out into architecture and was asked by a local worthy (of whom he did four portraits) to design a public market, to be named, after the worthy, Faneuil Hall. Finished in 1742, gutted by fire in 1761, the hall was remodeled in 1805 by Charles Bulfinch, who began in architecture as a gentleman-dilettante and ended in 1829 finishing off the Capitol in Washington, D.C. In between, he brought English style to America, and turned Boston from wood to stone with buildings like Massachusetts General Hospital, the State House, and Harvard University Hall. His devoted acolyte and copier was Asher Benjamin, who took Bulfinch's posh architecture out to the sticks (the small towns in Massachusetts and Vermont). His ideas really spread in 1806 when he published *The American Builder's Companion,* a how-to-do-it-in-Federal-style guide for carpenters all over the East. In the revised 1811 version, Benjamin included examples of the latest home technology: fireplaces including roasters and boilers, based on the work of fellow American Benjamin Thompson.

Thompson had fought for the Brits in the War of Independence (became Sir Benjamin), and was then made head of the Bavarian army (became Count Rumford), during which his research into gunpowder and cannon-boring got him into heat and then light, his two major obsessions in life. Minor obsessions were insulation, potatoes (he introduced them to Central Europe), healthy diet, coffee-making machines, the standard candle, and founding the Royal Institution in London for the education of craftsmen and the poor (in guess what: matters involving heat and light). In 1801 he hired as lecturer for the Royal Institution a young chemist who'd also written about guess what. Humphry Davy was soon laying 'em in the aisles (a thousand at a time) as London's most popular speaker. Davy went on to discover electrolysis, using it to isolate potassium and sodium, simultaneously inventing the miner's safety lamp with George Stephenson and in general becoming the appliance-of-science hero. Davy was such a big name that, in 1813, when Napoleon gave him a prize (even though at war with the Brits), both sides let Davy go to France to pick up the prize and schmooze with French nerds like Michel Chevreul—who was to out-fame even Davy (when Chevreul eventually died, France shut down for the day). Chevreul, another chemist, did everybody a favor with twin discoveries: how to make soap and candles cheaply. This, since he was a fats maven (he helped a young friend, Meges Mour-

tlers, to found Savannah. Two years later, they were joined by some more poor *and* oppressed settlers, Lutheran Protestants from seriously Catholic Salzburg, Austria, where, in 1731, they'd been given eight days to pack up and get out with whatever they could bundle together by the deadline. This, in spite of a Europe-wide agreement that you were supposed to give at least three years' grace as preparation to anybody who was to be chucked out for religious reasons. Not this time. Twenty thousand Salzburg Lutherans wrapped up what little they could carry and took off all over Europe to look for refuge in Protestant countries. One bunch was taken on by the English Society for the Propagation of Gospel Knowledge, put on a ship headed for Oglethorpe's paradise, and ended up founding New Ebenezer, Georgia.

The evil genius behind this Protestant diaspora was a nasty piece of Catholic work, the prince-archbishop of Salzburg, Leopold von Firmian, who spent the rest of his time adding some nice architectural features to Salzburg and hiring a fourth violinist for his court orchestra. Said fiddler, who also went by the name of Leopold, did not go down in history for his seventy symphonies (as I'm sure you'll agree) but instead became well known for his book on violin-playing, and then, shortly after 1756, universally famous forever, after the birth of a son who became universally known as Wolfgang Amadeus, who changed his father's life and that of most people. One of Leopold Mozart's favorite pieces was a set of twelve sonatas for harpsichord by an Italian composer and harpsichordist, who tried writing opera and failed, and who was known in England (where he had ended up by 1746) as Domenico Paradies. There's little else to be said about him, except he might have been copying the work and techniques of another, slightly better-known harpsichordist, Domenico Scarlatti. While in London, he taught Thomas Linley well enough for the latter to achieve the operatic success Paradies never had. Linley started out as a singing teacher at Bath, in the West of England, where the rich and famous went to take the spa waters and where there was a healthy cultural milieu. In 1774 Linley's fame had spread enough for him to be appointed oratorio director at Drury Lane Theater, London, where in 1775 his almost-first operatic attempt, *The Duenna,* premiered to rave reviews. A year later, he bought a quarter of the theater shares and carried on producing compositions that nobody listens to any more,

ries, invent margarine), because he was director of the Gobelins tapestry factory, and tapestry-making is all about how color behaves when it hits the animal fats in wool.

It is also why Chevreul went on to invent the Law of Simultaneous Contrast (why juxtaposed colors affect how you see them) and become the color guru for the color-juxtaposing Impressionists. Mystic, Romantic painter Philipp Runge was also doing the same experiments in Germany, painting (onto a wooden sphere) gradations of color between pairs of primary colors, plus black and white, and studying the psychological aspects of color perception ("harmonious" and "disharmonious" color pairs). In 1810 this stuff caught the polymath attention of Goethe, Germany's major intellectual behemoth at the time. Goethe's giant brain moved easily from novels, poems, and plays to philosophy, botany, optics, mining, architecture, horticulture, and comparative anatomy. His big thing was morphology: that organisms as we know them were different shapes deriving from the same archetypal, in-the-mind-of-nature "Ur-shape." In Weimar in 1798, while advisor to the local duke, Goethe recommended Friedrich von Schelling to a teaching job in nearby Jena, hotbed of Romanticism.

Schelling out-Ur'd even Goethe with his "Ur-matter": the fundamental substrate from which everything in nature was made. Change came, said Schelling, from the clash between opposites in nature and the resolution of that clash (thesis, antithesis, synthesis). This mystic gobbledygook became known as *Naturphilosophie* and excited science types everywhere, since it seemed to explain why the new mystery force—electricity—appeared to relate in some way to magnetism. In 1820 chemist Christian Schonbein met Schelling and became a believer. From then on, Schonbein was into polar opposites, like composition and decomposition—like what happened when you decomposed oxygen with electricity and got (he discovered) ozone. The nature-philosophical approach does not, however, seem to have caused his 1846 discovery of explosive gun cotton: dip cotton wool into fuming nitric and sulphuric acids, wash, and dry. All you had to do then, to achieve the biggest bang per buck that any military mind had ever heard, was to set light to the gun cotton. The line of arms-dealer customers went round the block. Then Schonbein took the next step (as they always say in those "toward the future" books about science): by dissolving gun cotton in ether, he came up with collodion. This hit an astonished public in 1871 thanks to the disappearing elephant scare when it was discovered the animal was "in danger of being hunted to extinction" *(New York Times).*

And also thanks to the efforts of an Albany, New York, printer, John Wesley Hyatt, who invented a substitute for potentially scarce ivory billiard balls (made from disappearing elephant tusk). Collodion

possibly because, as a contemporary noted, his work was "not distinguished by any striking marks of original genius" (i.e., he snitched a lot of other people's ideas). One of his fellow shareholders at the theater married Linley's daughter Elizabeth, wrote the libretto for *The Duenna,* and then became one of the all-time stage greats in his own right.

A few years earlier, Richard Brinsley Sheridan was a friend of the Linley family when young Liz decided to escape the unwelcome attentions of an over-eager army officer by fleeing to a convent in France. Sheridan agreed to help her get there safely. Once in France he proposed matrimony, the pair got married, and Sheridan returned to England to a great theater career. His résumé included *The Rivals* (1775) and *The School for Scandal* (1777). By 1780 Sheridan was good friends with major arbiters of literary taste and fashion, and then took the usual route out of showbiz into politics. He immediately became much better known for his parliamentary performances (one speech lasted two days) than the ones he had given at Drury Lane. Theatrically, he only ever put one foot seriously wrong, when in 1775 he agreed to stage a newly discovered "lost" Shakespeare play: *Vortigern.* You undoubtedly know the rest of the story. The play was one of a number of Shakespearean "autograph manuscript" fakes (including *Henry II,* and "Shakespeare's mortgage document"), produced by an enterprising young man named William Ireland, who made a niche for himself in the field of literary scams. The "newly-discovered-Shakespeare-texts" fraud got as far as receiving the seal of approval from a credulous Prince of Wales before it was blown (observant scholars noticed that Ireland tended to insert too many double consonants and finished every word with an "e"). Ireland confessed. And went on doing it, as well as writing copious quantities of novels, poems, and essays (none of which are worth putting even *this* book down for). One of his own-work efforts was a four-volume history of the county of Kent (1834), which kicked off a series in the same vein. One of which, on the county of Essex, was produced by Thomas Wright, the guy who first gave Anglo-Saxon and Medieval English to the general public with an unending stream of editions and translations of all the great texts.

In 1843 Wright and three pals set up the British Archeological Association. One of its first members was the unexciting Robert Willis, who was a Cambridge professor of

TRACK ONE

plus camphor plus alcohol plus heat gave Hyatt a moldable, then (when it cooled) solid, plastic material. This failed to solve the billiard-ball problem but in the end took over the world (in the form of collars, cuffs, brush handles, lampshades, and anything else you can think of) as celluloid. In 1889 a former bookkeeper and a buggy-whip manufacturer, who had teamed up to make and sell photographic equipment, found that gun cotton and alcohol also produced a thick, syrupy liquid, which dried to form a thin flexible shell. This became celluloid film negative, and it made the whip manufacturer and his bookkeeper partner (George Eastman) multimillionaires when it was rolled into cartridges that fit inside their new Kodak Brownie cameras.

END TRACK ONE

mechanics and, as gearing guru of the day, wrote such riveting stuff as *On the Teeth of Wheels.* Needless to say, Willis was skillful at carpentry. His magnum opus was *Principles of Mechanism,* a piece guaranteed to cure your insomnia. Willis took his mechanical approach to a study of the structure of cathedrals, taking them apart architecturally to see how they worked. He did this with Winchester, York, Norwich, Gloucester, Lichfield, Salisbury, and Wells and was planning to do the same for the Oxford and Cambridge colleges but, fortunately for posterity, died. Early on, in 1849, serving on a commission looking into the use of iron in railways bridges (how much would iron deflect under what weight?), he consulted fellow Cambridge professor G. G. Stokes, one of those mathematician-physicists who knew everything, and whose other research included how clouds form, what is fluorescence, how to calculate the shape of waves, why telegraph wires sing in the wind, and how light behaves. Apropos of this, one of Stokes's pals was a quiet chemist from the west of England by the name of William Bird Herapath, who became involved with dog urine and light. If you drop a solution of iodine into the urine of a dog which has been fed quinine (honestly, I'm not making this up), you get tiny crystals of iodosulfate of quinine. The supercolossal thing about these tiny crystals is that when you shine light through them they do stuff so interesting that William decided to give the crystals his own name: herapathite. You know about herapathite if you've ever wished you didn't have to squint on a sunny day, because herapathite will polarize light when all its crystals are similarly aligned.

END TRACK TWO

AND FINALLY . . .

Around 1930, an American inventor ground up herapathite crystals, put them in solution in a test tube, applied a magnetic field to make the crystals align, dipped a strip of celluloid into the solution, pulled it out, and dried it. The result was the first prototype of Edwin Land's crystals-between-layers-of-plastic, light-polarizing eye-savers we now call Polaroid sunglasses.

ELEVEN

1770: FALKLANDS WAR TO TELEVISION

The first Falklands War (of 1770) started the same way as the later version (of 1982): Spanish-speaking troops clobber heavily outnumbered English garrison, and take over. Unlike the second go-round, however, in 1770 the English lost and left the islands to Spain (while still claiming sovereignty). Things were complicated by the fact that the Spaniards had only recently bought the place from the French, who'd had people in East Falkland since 1764, the same year the Brits (led by Byron's grandfather) turned up and settled West Falkland. Each group thought it had the place to itself.

Falkland was fomented by the French secretary of state for foreign affairs, Choiseul, a canny diplomat and former soldier who'd risked life and limb on many battlefields (and by being the lover of Madame de Pompadour, the king's mistress). Choiseul enjoyed socking it to the Brits any time he got the chance, given that they had only recently relieved France of much of India and North America. For this reason, Choiseul was deeply into own-back rebuilding of the French navy, the plans for which were nixed when the unfortunate demise of his well-connected lady love brought the end of his career. Still, others persisted, in the form of his follow-on, former soldier and ladies' man, the Marshal de Castries, reformist minister for the navy from 1780, and the guy who helped organized French support without which American independence might not have happened. In time (1787) he, too, succumbed to bureaucratic in-fighting. Back in 1785 Castries had been introduced to a talented lad who wanted a life on the ocean wave and arranged a cadetship for him. This took young Marc Brunel to the Caribbean for six years. In 1792 the crew was paid off, and Brunel headed for New York, ending up that city's chief engineer. In 1799 he arrived in London with an idea he sold to the British navy for machine-made pulley blocks (those wooden things sailors run ropes through when they're hauling sail or guns or cargo or anything heavy). Brunel's invention would manufacture these objects ten times faster than people could, once he'd found Henry Maudslay, machine-tool guru, who used his new slide-rest lathe to make forty block-making machines, finally installed in 1806 at the naval yard in Portsmouth.

Later, Maudslay also developed something for measuring something teeny-weeny, thus increasing the precision with which metal could be turned and machines made. Chief helper in this meticulous task was Maudslay's draftsman, Joseph Clement, whose only love was precision. Clement was the guy who in 1828 invented the self-adjusting double-driver center chuck. And you read it here first. He also had a reputation for charging an arm and a leg (especially to American clients), which might be why he didn't last long on one project that was backed by the British government but which never came to fruition or profit (so Clement pulled out and so did the government). The idea was for a means of automating the calculation of stuff like longitude tables, actuarial tables, taxation, and all the sums that make the commercial world go round. The gizmo (that, in the end, never happened) was the "difference engine," designed to go to fifty decimal places and brainchild of Charles Babbage, math whiz and a man who never finished anything. This was possibly because he

TRACK TWO

In June 1770 Falkland events came to a head when five Spanish frigates carrying sixteen hundred men and mucho firepower sailed into Port Egmont harbor, West Falkland, and invited the commander of the single British ship there to look them over. Outflanked and outnumbered, Captain George Farmer made a show of firing guns and then surrendered. After sending a letter to his Spanish counterpart saying the place would remain British, Farmer sailed for England and, given the situation, no blame. By 1773 he was on his way to the East Indies in command of *Seahorse.* Onboard was an unknown young midshipman named Horatio Nelson and a young able seaman, Thomas Troubridge. Troubridge's career was slow and steady until 1797 when he took up once again with his erstwhile shipmate, now Admiral Nelson and a national hero. For two years, they chased the French fleets around the Mediterranean. In 1799 Troubridge was limbering up for what was to be the glorious Battle of the Nile (Nelson won, Napoleon lost) when his ship ran aground and he had to miss all the fun. However, as Nelson's pal he received a gold medal and a baronetcy just for sitting on the rocks. Win some, win some. In next to no time, Troubridge was a junior lord of the admiralty under Earl St. Vincent, another dashing naval hero who'd won the Battle of Cape St. Vincent and received a vote of thanks from the secretary of war and the freedom of most cities in Britain. Vincent was a tough cookie: hanged mutineers, forbade officers from sleeping anywhere but aboard, and generally ran tight ships—afloat or ashore.

This proved to be the case with his government commission into naval corruption and theft (which in 1805 he revealed was rife and was costing the government tens of millions a year). Criminal negligence, said his report, went all the way to the top, as in: Lord Melville, secretary of war and treasurer of the navy. Everything hit the parliamentary fan. Back in 1872, Melville had led the attack on the man in charge of British India and brought him to trial for mismanagement. Now it was Melville's turn. Offered the choice of either impeachment for "high crimes and misdemeanors" or a criminal trial, he sensibly opted for impeachment. The parliamentary proceedings lasted fifteen days, and at the end Melville was acquitted of the most serious charge (he'd let his assistant siphon money out of the Bank of England).

was interested in everything: cryptanalysis, probability, geophysics, astronomy, altimetry, ophthalmoscopy, statistical linguistics, meteorology, actuarial science, lighthouse technology, and the use of tree rings as historical climactic records. He also hated organ grinders. And military amateurs who played at science.

In this case, he meant one Captain Edward Sabine. Accused by Babbage, in his report on the chaotic state of Royal Society affairs, of being a "charlatan" who got his data wrong (in one case, by a factor of ten). The data in question were magnetic variation figures that Sabine spent most of his life collecting (or organizing the collection of) at stations around the world. Sabine's Great Magnetic Crusade was also meant to prove his theory that there were two magnetic poles in each hemisphere. He also managed to correlate the sunspot cycle with magnetic storms. But most of his time was spent maneuvering with bureaucrats to secure grants for his pet projects. Two of which were Greenland birds and Eskimos. In 1836 he met Ben Franklin's great-grandson Alexander Bache, another magnetic freak. Bache was a West Point graduate who'd recently been blowing up steam boilers for the U.S. government (to see why boilers sometimes blew up). He went on doing these officially explosive things, meantime becoming one of the senior people working out the delicate relationship between science and government.

In 1843 Bache became superintendent of the American Coast Survey, which was to attract more federal funding (and hire more scientists) than any other project of the day. In 1843 one of the (nonscientific) hirelings was one Isaac Ingalls Stevens, another West Point man, who'd just come back from the Mexican War. He then campaigned for Pierce, the winner of the next presidential election, which got him the governorship of the newly created Washington Territory. And with the area's future (and his own career) in mind, he jumped at the chance to lead the survey of a possible northern route for the transcontinental railroad (four routes were being proposed). The northern route was to go from St. Paul, Minnesota, to Puget Sound, and Stevens decided to split the survey into two teams: one to go west from St. Paul and the other heading east from Port Vancouver. To lead the second team, Stevens chose George McClellan, a fellow officer from his Mexican War days. McClellan's job was to find a pass, through the Cascades, that would take the railroad. In 1853, at Fort Colville near the Canadian border, he met Stevens coming the other way and reported that he'd seen only two reasonable passes but that neither would work because of the snow levels in winter. With so much at stake, Stevens sent his own men to double-check. They found five workable passes (with acceptable snow levels) that McClellan had missed or had been too lazy to check.

But the acquittal majority was so small that it was a guilty verdict in all but name. It ended his career. Point man for the government in the affair was Sam Whitbread, brewery owner and giver of more speeches in Parliament than anybody else. Sam was for liberal reform, free education, badges for the really poor (no badges for fakes), and was pro-American over the War of 1812. That year he met Princess Caroline, wife of Prince George (heir to the throne) and was smitten. Caroline was unusual in that she almost never washed and indulged in "indecent conversation." She also had numerous lovers, possibly because her husband was keeping at least two mistresses and their marriage back in 1795 had lasted all of three weeks before informal separation. But the public loved her. In 1814, after years of royal isolation and ill treatment, she left for Italy and another lover. In 1820 it became George's turn to be king, so she headed back to be crowned queen. No deal. They even shut Westminster Abbey's doors. This was all too much for George, and a "trial" by Parliament was arranged (George wanted grounds for divorce). In 1821 Caroline died.

Her defense counsel in Parliament had been the ambitious Henry Brougham, who made his name with the case and went on to become lord chancellor (top lawyer), founder of London University and the Society for the Diffusion of Useful Knowledge, and inventor of the brougham (less a carriage, more a garden chair on wheels). He also built a house in Cannes and kind of started the South-of-France craze. Brougham also wrote for the *Edinburgh Review* on everything, best of which was literary criticism that generally went for the throat. His editor was the mild-mannered Macvey Napier, who had edited the seventh edition of the *Encyclopedia Britannica,* took over the *Edinburgh Review* in 1829, and was good at handling prima-donna contributors, as a result of which he got some big-name bylines. One of these, Sir William Hamilton, wrote a series of articles that earned him an international reputation. Hamilton was the guy who (wait for this) first quantified the predicate. You get a vague feel for what that involves from his row with French philosopher De Morgan, who argued that Hamilton was wrong because: " 'All A are some B, and some B are all A' can only be true if, and only if, all A are all B." The two of them also had a set-to when De Morgan sent a paper to eminent professor William Whewell, then got a paper from Hamilton, then recalled his paper from Whewell and rewrote

In the event, the matter became academic when Secretary of War Jefferson Davis chose a southern route. No surprise, given that he was a Southerner. Through the run-up to the Civil War, although a powerful advocate for states' rights and a defender of slavery, Davis played the conciliatory role. When the war broke out and he was elected president of the Confederate States, he proceeded to use his powers to an unprecedented extent: military draft, a powerful central government, no habeas corpus, and control of rail and shipping. All this managed to eke out the South's meager resources, but in the end the sheer industrial weight of the North would prevail. In 1863 Davis wrote to Pope Pius IX praying for peace, and the pope wrote back, thus, Davis claimed, implicitly recognizing the sovereignty of the South. Two years later, when Davis was in a postwar prison, Pius wrote again, this time sending a self-portrait showing himself meaningfully touching a crown of thorns. Like Davis, Pius had suffered a lost war. After early attempts at political liberalization in the papal states had merely served to enflame the radicals, Pius found himself trapped by a mob in Rome in 1848 and was only kept on the throne by the presence of a French army. By 1860 the drive to unify the country was inevitable, thanks in large part to a charismatic adventurer in a red shirt named Giuseppe Garibaldi.

After twelve years' exile in South America, a year in the United States, and a period of revolutionary skirmishing back home, in 1860 Garibaldi found himself, and a thousand other red shirts, heading for Rome and a united Italy. By 1870 Pius's territory was reduced to a tiny Vatican state. During the final moments of the struggle, Garibaldi was almost aided by the (late) arrival of a group of 810 volunteers from Britain, where the government had turned a blind eye to the adventure so long as the volunteers' uniforms were for the "purpose of recognizing each other" and their guns were for the "purpose of self-defense." Organizer of this brave band was socialist George Holyoake, the radical's radical (he tested bombs for other radicals). George invented the word "secularism" and at one point got jailed for blasphemy. He agitated for such insanities as freedom of the press and electoral reform and, like so many of his ilk, visited the United States twice to see a liberal future in action. His 1884 visit was enlivened by contact with the Colony Aid Association of New York, whose (doomed) ambition was to settle the West according to moral principles. A major figure in this well-meaning task (and who funded Holyoake while in the United States) was Elizabeth Thompson. Started poor, she was working at age nine, later married a rich man twenty-three years older, and when he died gave much of his money to good causes (the Colony project, a woman's medical hospital, a telescope for Vassar). In the 1880s she turned her benefactions

it. "Plagiarism," said Hamilton. "Minor additions," said De Morgan.

As far as I know, Whewell kept out of it all. His field was history and philosophy of science, and studies so general as to be considered dabbling by today's doctoral standards. Apart from writing sermons, poetry, and translations from Greek and German, Whewell studied crystallography, geology, theology, political economy, and architecture. And coined the terms "physicist," "scientist," "ion," "cathode," and "anode." And did the best work on tides since Newton. And was also vice chancellor of Cambridge, whose curriculum he dragged kicking and screaming into the nineteenth century (where have we heard *that* before). In 1832 Whewell also persuaded a number of worthies to get interested in the new statistics gobbledygook. One of these people was, logically, a Belgian astronomer by the name of Quetelet. He was the person who conceived of the "average person," as part of his work on "social physics" (a.k.a. sociology), and dreamed up the "statistically meaningful sample" (of which you may have been one, at some time). Quetelet's approach derived from the kind of math astronomers use to calculate stuff like orbits (I think) and other such constantly changing movements—like the weather, on which subject Quetelet was also a mover and shaker, convening the first international conference. One of his meteorological pals, Francis Galton, identified and named anticyclones.

Galton was obsessive about samples of all kinds, including surveys to find the site of the most beautiful British women (Aberdeen), the efficacy of prayer (not much), and the average body weight of three generations of British aristocrats. Once his cousin Darwin's blockbuster was published, Galton was instantly into heredity and eugenics and all that (later misused as Nazi racial-purity gibberish). Darwin had his own health problems to think about. He dealt with them at the fashionable water-cure center run by Dr. John Gully, a medical smoothie (formal education at the cutting-edge Paris hospitals) who attracted celebs like Tennyson and Dickens to his hydrotherapy establishment and got endorsements from hardheads like Florence Nightingale. Gully's career took a dive late in life when he was mixed up, in some ambiguous way, with a woman who poisoned her husband. But while still respectable, he became physician to the greatest medium (or cheat) who ever lived, the American D. D. Home. From an early age, when Home was

toward science and the good it could bring society. In 1889 her Science Fund dumped many dollars on two inseparable German noodlers, Julius Elster and Hand Geitel, who were doing important fundamental work in atmospheric electricity, radioactivity, and looking at how some minerals reacted to light by giving off electrons (electricity). In 1893 the pair produced their device. You shone a light at it (even ultraviolet light) and it responded by producing current that varied according to the intensity of the light. By 1899 they had a fully practical version; they called it a "photoelectric cell."

END TRACK ONE

around, tables tilted, knockings were heard, and lights flickered.

In 1855 he arrived in England and was an overnight sensation. Soon he was doing trances for the crowned heads of Europe (this resulted in his marriage to the cousin of the czar of Russia and a giant diamond ring presented by the man himself). Anybody who was anybody (except Robert Browning) was a believer. Home was even received by the pope. In 1871 he passed the final test, being investigated by one of the greatest scientists alive, physicist Sir William Crookes. Crookes, after measuring the experimental ambience (temperature, weight of all objects in the room, full daylight, etc.) reported that Home then had, indeed, lifted off the ground, moved objects at a distance, caused automatic writing, and done other magic. Mind you, Crookes wasn't entirely unbiased. He was, for instance, infatuated by a ghost named Katie King and had seen his own share of accordions playing in midair. In spite of this, in 1878 he still found time for heavy-duty science such as experimentations with extremely low-pressure gas in a vacuum tube containing two electrodes. When a high voltage was applied, a glow was caused (by the stream of electrons) that could be deflected by a magnet.

In 1897 a German named Braun (who'd written his doctoral dissertation on vibrating elastic strings and rods) picked up on Crookes's idea. By applying an alternating electromagnetic current to the stream of electrons, he was able to make patterns with the glow point made by the electron beam on a fluorescent section of tube. He called his invention an oscilloscope, and it instantly became a tool no researcher went without.

END TRACK TWO

AND FINALLY . . .

In 1908 a Brit, Alan Swinton, described how you could focus an image onto a mosaic of photocells, use a cathode ray to pick up the varying intensity of current the cells gave off, send this varying-intensity current to a cathode in an oscilloscope, and use its electron beam to trace out the original image. This neat trick would eventually become known as television.

TWELVE

1724: STONE AGE BOY TO PHOTOCOPIER

One summer morning in 1724, in a field outside the northern German town of Hamelin, somebody came across a naked young boy. The child (who became known as "Wild Peter") appeared to be in a feral state. He walked on all fours, climbed trees like a monkey, sat on his haunches, ate only raw vegetables and grass, and spoke no language. The Age of Enlightenment took to him with all the abandon of an alcoholic in a brewery. Was he an ancestor of modern man? If he could be taught to speak, would he reveal how humans, unspoilt by civilization, viewed the world? What did his condition say about the nature/nurture argument?

TRACK ONE

In February 1726, "Wild Peter" was sent to England as a present for the Princess of Wales, who spent time failing to teach him to speak, and then passed him on to her doctor, the absentminded John Arbuthnot, who made no better headway. The boy was then returned to Germany, where he lived to a ripe old age, as silent as he had begun. Arbuthnot's mathematical interests then led him to study the records of christenings in London between 1629 and 1710 and to reveal to an astonished world that the near-equivalence of male-female births was clear proof of Providential Involvement in daily existence. Nature was producing a boy for each girl, plus a few extra boys because men tended to get killed more often than women—the first example of a mathematical statistical inference.

Arbuthnot's social position bumped him into lit. and wit, and he was soon a writing member of the Scriblerus Club, dedicated to pricking balloons of all pompous kinds. Fellow member (and great pal) was archsatirist Alexander Pope. Satire, you feel, was Pope's armor against the world, being deformed and four feet six inches. In 1711 Pope hit the big time on his first attempt, with a piece rivetingly titled *An Essay on Criticism,* a poetic review of the whole of literature, containing aphorisms as far as the eye could see, including some we still know so well that you can fill in the blanks: "Fools rush in . . . ," "A little learning . . . ," "To err is human . . ." In no time at all, Pope was part of the sardonic set, which included sharp tongues like Addison, Congreve, Steele, and Swift. The kind of people you didn't want as your enemies in an age when reputation was make-or-break, as Pope was to prove, in a lifetime of cutting remarks. The other kind of cuttings Pope was crazy for were those in his garden in Twickenham, just outside London. Here, he lived next to the local vicar (of St. Mary's), a quiet, self-educated bumbler whose Thames-side potterings turned him into one of England's greatest scientists and the inspiration to such brains as Priestley (discovered oxygen), Cavendish (discovered hydrogen), and Black (discovered latent heat).

Though he was certainly no fool, Stephen Hales and sap are inextricably linked because that's what Hales cared about most: the question of why sap rose and by how much. To which end, he would stick glass tubes into anything vegetable and watch how far up the sap went, by night, day, hot, cold, and season. He discovered that the pressure behind the sap was seven times the pressure behind that of a dog's blood. (He knew about the dog thanks to similar experiments on more than one unfortunate). Results convinced him that by the time blood got to the capillaries, it wasn't forceful enough to move muscles (the contemporary theory). There was only one other possi-

TRACK TWO

Well before "Wild Peter" finally died, in 1775 his case was studied by Johann Blumenbach, the man responsible for the term (still in use on police blotters until recently) "Caucasian." It was Blumenbach who divided the human race into five types (Caucasian, Malayan, Ethiopian, American, and Mongolian), according to the shape of their skulls. Of which you got the most accurate view by placing the skull between your feet, thus seeing it in the "Blumenbach View," one of the various experimental techniques developed by Blumenbach and which marked the start of the modern science of anthropology. Blumenbach's main aim was to dispel the myths surrounding human origins. His masterly *On the Natural Varieties of Mankind* (an instant best-seller) dismissed ideas of men with no heads, with eyes on their shoulders, with dog heads, or twenty feet tall. He also dispelled the "Wild Peter" myth, noting that, when discovered, the boy had scraps of clothing on him and that his pale thighs and sunburnt calves meant he'd worn breeches and was probably somebody's dumb offspring.

One of Blumenbach's pals at Göttingen University was another myth collector, by the name of Heyne. After a typical classical-scholar start (no money, odd jobs cataloging books in aristocrats' libraries), in 1763 Heyne was given the chair of poetry and rhetoric at Göttingen and set about putting the Greek myths under scientific scrutiny. Stimulated by the latest information about Native American oral traditions, he came up with the (now-accepted) idea that myths were ancient explanations of how the world worked and that to understand prehistoric peoples you had to understand their worldviews. In general, to appreciate the art of ancient societies, you needed to know how those societies worked.

Earlier, in 1753 (in his library period) Heyne had met, and passed on these thoughts to, Johann Winckelmann, who was doing the same book-sorting job for some other count. In the course of this dusty work Winckelmann then came across Greek art and went promptly bonkers about it. He ended up in Rome, doing inventories of various collections, visiting archeological excavations, and mixing with the gay art community. In 1764 he produced the book that would establish the field of art history, titled *History of the Art of Antiquity*. There was not a word in it about Heyne.

ble mechanism: nerves (this opened the way for Galvani's work on animal electricity). Hales's discovery of air bubbles in sap led him to investigate plant (and then general) respiration and, from there, noxious-air ventilators in prisons and ships. The end result of all this heavy-breathing work was modern respiratory medicine. Not bad for a bumbling vicar, right? By 1753 Hales had royal admirers (the Prince and Princess of Wales), which meant he would be one of the founders of the Society of Arts.

This was the idea of otherwise-unknown William Shipley, who wanted "to embolden enterprise, to enlarge science, to refine art, to improve our manufactures and to extend our commerce"—in other words: to make England rich and powerful. Small wonder an early aim of the Society of Arts was to foster tree planting (fifty million by 1821, many of them still there today) and, later, to set up plantations in the Caribbean to grow (for profit) stuff like cotton, breadfruit, and nutmeg. By this time the Brits were running the place, having won a recent war and taken the islands from the French (and most important, the island sugar: the economic equivalent of modern-day petroleum). This explained why a professional soldier was in charge of the Caribbean island of St. Vincent and in 1763 set up the first botanical garden in the Western Hemisphere.

General Robert Melville (also a member of the S. of A.) got his governorship as the usual "grateful government" golden handshake after a lifetime of military service. In 1759 this included inventing a nifty little eight-inch howitzer, which could be fired from a ship (creating a carronade) and spoil any nearby Frenchman's day (as happened in 1782 when Admiral Rodney did it to Admiral De Grasse). From 1779 the gun was being made at the John Roebuck ironworks on the River Carron, in Scotland. By then Roebuck had gone on to greater things. A doctor with an interest in chemistry, he'd first made a fortune in sulfuric acid (used for bleaching), parlayed that into an ironworks, and used the profit to pay for coal mines and a leased-ducal-palace lifestyle. At this point his luck ran out. The mines flooded faster than he could drain them. In desperation he financed James Watt (whose steam engine was a pump long before it became a locomotive), but Watt took too long, so Roebuck had to sell his shares in everything and work as the mines' manager. In 1794 he died, seven years before a discovery that might have made all the difference to his fortunes.

In 1801 David Mushet, a self-effacing local expert on anything to do with iron, discovered immense (that is, cheap) deposits of blackband ironstone. This would provide iron for as long as anybody could imagine, except for the fact that it took more money to pay for the fuel to smelt blackband iron than you'd make by then selling the iron.

Early in his Roman sojourn, Winckelmann's money ran out and he was offered a job by his friend Alessandro Albani, with whom he'd been digging. Albani was the pope's nephew, a cardinal's brother, a cardinal himself, and a prince. Life was not tough. By this time, he was also developing a reputation as an art dealer (not difficult, in his position, to find stuff for well-heeled foreign nobility with a yen for Roman anything). His own collection became so big that he had to build a villa for it. Albani's social position gave him an entrée to every social whirl—useful when it came to helping his other excavating pal, Baron von Stosch, whose specialties were gems, coins, and depravity. Stosch never quite became the accredited diplomat he wanted to be, but his antiquarian expertise was an excellent cover for the next best thing: spying. Which he did for the British (with inside information from his know-everything pal, Albani). From Rome to London, for-your-eyes-only invisible-ink dispatches went, once a week, written by "John Walton"—until Stosch's cover was accidentally blown in 1731. One night not long afterward, three gunmen stopped him in the street and politely suggested he get out of town. So he moved to Florence for the rest of his life.

The reason for all this cloak-and-dagger hugger-mugger was that earlier, in 1715, the recently imported (German) royal family of Britain was shaken when a would-be coup by the previous (Scottish) royal family of Britain nearly succeeded. Failed coup leader, James Stuart (known as the "Pretender" from his pretensions to the throne), hightailed it to Rome, papal protection, and (one day, who knew) a comeback—hence Stosch and all that undercovery. In reality, the coup was never a serious threat. James was a drunken slob (supported by nobody other than the perennially anti-British French), who went around behaving like James III of England (which he would have been if things had gone differently). His son Charles wasn't much better. He started well enough (starstruck followers even named him "Bonnie Prince Charlie"), but when his uprising failed in 1745, he went the way of his old man, living off the courts of Europe and then moving to Rome to sponge off the pope, and calling himself Charles III of England (which he would have been if things had gone differently). He used the title at his wedding with poor Louise von Stolberg, who had thought her 1772 proxy marriage was to a dashing young cavalier (his PR handout said so), instead of what turned

So there it was left for twenty-six years, until the manager of the Glasgow gasworks, one James Beaumont Neilson, looking for a new market for his gas (and, anyway, interested in everything that burned), calculated that if he blasted hot air (preheated by passing it through giant gas-fire rings) into a furnace, it'd do money-making things. For example, it could produce three times the iron for the same amount of coal—just what those cheap, blackband ironstone deposits were waiting for. Before you could say "Industrial Revolution," the Lowland Scots were having one. They never looked back.

Neilson's idea worked out (as is always the case) because hot blast came in the right place at the right time. And so did Charles Macintosh, the man who gave Neilson the money to install the system and who then became his partner. Macintosh already had a deal with Neilson to remove the throwaway by-product of the gasworks: coal tar. From this, Macintosh was distilling out ammonia (cheaper than the previous source, urine) for use in his dyeing works. It was during this malodorous operation that Macintosh found that he was also by-producing naphtha, which is a chemical that will liquefy rubber. So he did just that, in 1822 spreading the liquid rubber between sheets of cloth and giving the world a raincoat (Brits still call the garment a macintosh). Minor problems emerged. In winter the rubber stiffened and cracked; in summer it became sticky and smelled awful. Solving these problems became Charles Goodyear's life's work. Charles was no scientist, so by trial and error he came to vulcanization of the rubber (mix with sulphur and lead oxide, and heat), which turned rubber into an instant economic success—except for Charles. Patent piracy and Charles's ingenuousness led him all too frequently to debtor's prison (he was in a French jail the day he was awarded the Legion of Honor), and in the end he died in hock. But he'd made his mark. In 1898 another American (Frank Sieberling, scion of the Empire Mower and Reaper family) set up business in rubber and named his company Goodyear in honor of Charles. By 1916 it was the biggest in the world, thanks above all to the explosive growth of automobile manufacture and the need for tires. It made Akron.

In 1889 or 1890, the Goodyear company was approached with an idea by an ex-cocaine-and-morphine addict, William Halsted, a surgeon who had experimented on himself with the drugs as part of his attempt to discover neuroregional anesthesia (nerve blocking). The work earned him his habit and the gold medal of the American Dental Association. In the end he stopped cold turkey, got on with his life, and by the time he contacted Goodyear was professor of surgery at Johns Hopkins. His head nurse (later wife) was complaining of dermatitis contracted through handling antiseptics. Goodyear provided the surgical gloves Halsted had requested and revolutionized medical

out to be a sot and a lousy cello player. By 1773 Louise was looking elsewhere for solace, and in 1776 she found it in the Italian poet Alfieri. In 1784 she and Charlie separated, and she and Alfieri moved to Florence, where he became a workaholic. By 1793 Louise was on the lookout once more. She didn't have far to look. The sympathetic French refugee painter Francois-Xavier Fabre was living with them at the time.

A couple of years later, the English Lord Holland was in town, and Fabre was doing his portrait. In tune with the wife-swapping milieu, Holland had the hots for Elizabeth Lady Webster (the spouse of a friend), who was alone in Florence with one of her girlfriends and, so to speak, getting notches on her gun. One of whom soon became Holland. Faster than you can say "divorced," she was. They then married (a scandal even in those no-holds-barred times) and returned to London. There she set up at Holland House, which for thirty years then became the place to go if you wanted targets on which to practice your razor-sharp wit. It was said pharmacists did well selling feel-better pills for people who'd been savaged at Holland House during dinner-table repartee that was little short of fatal—unless, of course, you'd survived much worse. This was the case of one of Elizabeth's favorite guests, the French diplomat and general charmer Talleyrand. After finessing his way through the shark-infested waters of the French Revolution, then cozying up to Napoleon, then to the royal follow-ons, Talleyrand made it safely to old age and to a London ambassadorship.

Early on, a sweet tooth had taken him to a famous Parisian patisserie, where he fell for the extravagant confections of a young pastry cook named Carême and lured him away to become his chef for ten years. After which, Carême moved on to do the same trick for the czar, the British prince regent, and assorted luminaries, including of course the Rothschilds. At the height of his Paris career, Carême ran a restaurant and attracted a customer who had studied in Bologna (center of the food universe) and was such a gastronome he put food in all his operas—and almost anything else you can think of. He boasted: "Give me a laundry list and I'll set it to music." It was for Rossini that Carême is said to have invented *tournedos Rossini* (sauté in butter slices of beef fillet, cover with croutons, top with a slice of sautéed foie gras and truffles, drench with pan juices

practice—once, that is, they'd found an easier way to don the gloves. It was easier with a powder (one of whose ingredients was lycopodium spores), which was in general use in hospitals by the 1920s. Until (in the 1930s) they found it had toxic properties undesirable in proximity to an open wound. That put the kibosh on lycopodium—for the time being.

END TRACK ONE

TRACK TWO

diluted with Madeira and *demi-glace).* Rossini loved opera, good food, and women in that order—and did well in all categories, with one noisy exception. In 1822 his *Cenerentola (Cinderella)* got clobbered by the French critics as a takeoff of the 1810 production *Cendrillon (Cinderella)* by a minor composer known as "the Maltese" after his birthplace. The injured party in question, Nicolas Isouard, was a protégé of the Knights of Malta and an early exponent of comic opera. *Cendrillon* was the biggest success of the time at the Opera Comique, so you have to wonder whom Rossini thought he was fooling. In 1822 Isouard's other hit was *Aladdin, or The Wonderful Lamp,* in which he aptly introduced the first use of gaslight in the theater.

By this time the new wonder illumination was being used in cities all over Europe and the United States, and the gasworks were generating tons of a stinking by-product (coal tar), which they then chucked into any convenient river or pond. Given that anything which comes free is going to interest people with money in mind, by 1839 profit-conscious chemists had boiled, dissolved, and distilled coal tar and found naphtha, ammonia, creosote, and an antiseptic. In 1865, this last medicinal by-product of the by-product had fired the imagination of a young Brit chemist, William Perkin, who believed that the evil-smelling black sludge contained the secret of artificial quinine. Alas, the muck Perkin finally produced was not artificial quinine. But it was the world's first artificial aniline dye: mauve. This made the fashion industry happy and Perkin a rich man. Encouraged, Perkin dived back in. This time (1869) his coal-tar experiments succeeded in deriving industrially significant amounts of anthracene.

END TRACK TWO

AND FINALLY . . .

In 1938 American Chester F. Carlson used anthracene to form a thin photosensitive layer on a zinc plate. When he exposed the plate to an illuminated image, the layer charged only where the image was. He then covered this charged area with lycopodium powder and carbon black, laid a sheet of charged paper on top, and the powder stuck to the paper. Heating the paper fused the powder to the paper in the shape of the image. Today, we call the process photocopying.

THIRTEEN

1745: LEYDEN JAR TO CLINGWRAP

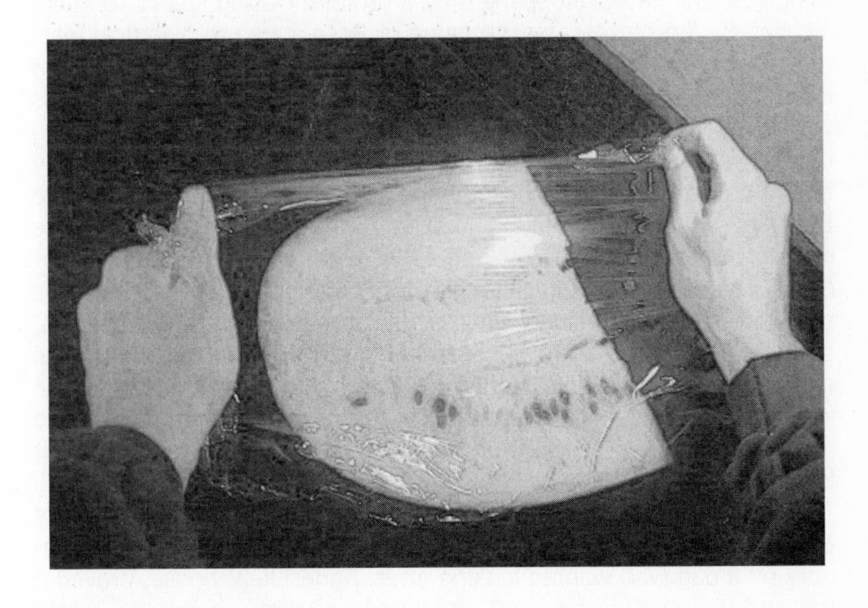

Sometimes an idea is in the air and is picked up by more than one inventor. Such was the case in 1745 Germany and 1746 Holland with the following: almost fill a glass container, stoppered by a cork, which is pierced by a wire that touches the water surface. While holding the glass, bring a static-electricity source in contact with the external end of the wire. Remove the electricity source. While still holding the glass, touch the wire and get a shock. This shock is produced by the electricity now stored in the "Leyden jar."

TRACK ONE

The first person to invent the Leyden jar (after attending Leyden University, the M.I.T. of the period) was Ewald von Kleist, a German bumbler who then made a stupid mistake. When doing his show-and-tell, he forgot to mention that you have to be holding the jar while it's charging and also when you touch the wire with your other hand. Otherwise, nothing shocking happens. That is what happened (or rather, didn't) when others tried the trick. Nobody got a charge out of something that did nothing, so it was "come-back-when-fixed" time. Poor old Kleist began his rise to anonymity because his interest in things electric had been sparked by a professor using sparks to ignite alcohol. Said professor, by the name of Bose, came to electrical matters after writing about eclipses and mistakes made by doctors. When Bose thought up his scintillating electrical displays, he, too, got things wrong. In 1752, at the Lutheran University of Wittenberg (founded to counteract the evils of Rome), he proceeded to send a handsomely bound copy of his research, together with an effusive letter, to Pope Benedict XIV. Mind you, this was no run-of-the-mill pontiff. Before getting the top job, Benedict had spent years being the devil's advocate, checking into applications for canonization. This process required that the candidate display "heroic virtue" and four miracles, so Benedict's task was to put wanna-be saints under the theological microscope to look for evidence of fakery. It helped that he was a keen microscopist—and mathematician and general science buff. So he knew what was going on in the work of toys-for-boys—and, as it happened, girls.

In one case, it was Signorina Maria Gaetana Agnesi, who by the age of nine was fluent in seven languages. Withheld from the nunnery by a dad who wanted looking after, Agnesi kept house, prayed a lot, and did good works—and, at the age of thirty-something, wrote the first textbook on calculus. As a result of which Benedict offered her the chair of math at Bologna University. Agnesi preferred good works, private study, and the discovery of a cubic curve now called the curve of Agnesi. Math-oriented readers will be impressed—as was the pope, and an Englishman, John Colson, who translated her book not long before 1760. No slouch himself, Colson had published commentaries on Newton and was Lucasian professor at Cambridge—pretty good for a self-taught, no-degree amateur. One of his pals was an overqualified cleric named William Cole, who spent his life moving from the ruins of one disaster to another (his home fell down, floods ruined his property) and never stopped worrying about money. Cole was an antiquarian whose great claim to fame was his collection of one hundred volumes of manuscript he had written over decades: his

TRACK TWO

The other inventor of the Leyden jar did it by accident. In 1746 Petrus van Musschenbroek, professor at Leyden, was charging up a bottle of water via a wire going through the cork, when his assistant (holding the bottle) touched the wire and discovered what people mean by "getting a charge out of your work." After which Petrus explained in every detail what had happened, so people could repeat it, and in no time stored electricity was being used for everything from quack cures to causing a line of hand-holding monks to leap in the air. Petrus became famous and people turned up to make contact. One such being a 1754 visiting English engineer, John Smeaton, who was over in Holland to look at dikes, windmills, canals, and other stuff that a country's good at when it's wet and waterlogged. A year later, the wooden Eddystone lighthouse (near Plymouth, England) burned down. The earlier wooden one had blown down. Smeaton had the idea that dovetailed blocks of stone might fare better in wind and fire. So by 1759 he'd built it, and the new lighthouse stood for another 118 years. Smeaton progressed to bridges, canals, and harbors, and in 1771 started a club for his fellow professionals, which eventually became the Institute of Civil Engineers. Shortly thereafter, Smeaton happened to be repairing a Newcomen steam pump and came into contact with Matthew Boulton, who knew a bit about such things because his partner James Watt was about to render the repair of (now-out-of-date) Newcomen steam pumps a waste of time.

Boulton had started out as a maker of metal shoe buckles and expanded into snuffboxes, watch chains, and metal copies of art objects of all kinds. By 1780, after he'd taken over a share of Watt's patent (in return for settling Watt's debts) and the new steam engine was being leased everywhere, both men were coining it. In Boulton's case, this also meant his steam-powered, coin-stamping machines, which created currency for various governments. That year Boulton and Watt briefly hired draftsman William Playfair. By 1789 he had become agent for would-be emigrants to Ohio, then joined in storming the Paris Bastille, then failed to invent a successful semaphore, then spent his life as a starving scribbler doing pamphlets and translations. This, in a sense, was why his big brother John would become famous. John was an altogether more serious person. Pro-

thoughts on everything, plus notes on every book he ever read, plus drawings of churches, coats of arms, statues, and other higgledy-piggledy memorabilia.

In 1765 Cole went on a trip to France (to look for somewhere cheaper to settle down but didn't find it) with an old Eton school friend, Horace Walpole. Walpole (son of the Prime Minister, an earl, and loaded) invented two genres: one architectural, the other literary. First, in 1747, when he turned his small house outside London into a tiny, over-the-top, pseudomedieval castle, that set the style for Strawberry Hill Gothic. And second, when, waking from a nightmare in the above bijou residence, he wrote his dream down (replete with ghosts, unnamed terrors, and the period equivalent of the car chase, where the heroine is pursued, by an unknown presence, along a subterranean corridor). *The Castle of Otranto* was the first Gothic novel and the precursor of every bloodcurdler.

On the 1765 French trip with Cole, Walpole stopped off in Paris and fell for a blind, ill-tempered, bored-out-of-her-skull salon keeper, Mme du Deffand, and wrote to her frequently for the rest of her life. Deffand's salon was where you were seen if you were an author, philosopher, statesman, scientist, artist, or anybody with a witty remark to make (so stellar were the regulars that Voltaire was no big deal). Deffand ran the place because she had spent her entire life stultified by the vapid existence of the average French aristocrat ("All conditions seem to me equally unhappy, from the angel to the oyster"). The latest fashions in everything from plays to pedagogy were dissected and dismissed. Deffand started to go blind in 1752 and persuaded the gorgeous young Julie de Lespinasse (illegitimate daughter of an aristocrat and possibly Deffand's niece) to move in and be her companion. For ten years the pair managed the salon, with Julie getting better and better at the chat, until she began to do her own thing, in her own room, in the afternoons when Deffand was asleep. Because of this, a major row erupted in 1764 and Julie left (taking some big names), set up on her own with hand-me-down furniture and money from rich friends, and proceeded to out-Deffand Deffand.

In 1772, after an unrequited passion for a Spanish noble, Julie fell for a real cad: Joseph Count de Guibert. He was conceited, arrogant, and treated her badly (he kept a mistress all through his time with the lovely Julie and, unknown to both women, spent all his spare time arranging his eventual marriage with a nubile young heiress). By 1773 Colonel Guibert's name was on all salon agendas, thanks to his *Essay on Tactics* which took the military and political establishments by storm. In the *Essay*, he presented the novel concept that armies should be conscript and revolutionary, winning by sheer numbers and ruthless no-quarter force, the expression not of professional traditions

fessor of mathematics and natural philosophy (a.k.a. general science) at Edinburgh University and editor of the *Scottish Royal Society Journal,* Playfair knew VIPs like the economics guru Adam Smith and the geologist James Hutton before anybody else heard of them. This might never have happened in the latter's case, had it not been for Playfair. Hutton was a farmer turned geologist (before the discipline existed), and years of rambling and tapping the Highlands of Scotland with his little hammer persuaded him that the processes of erosion had likely been the same since Year One. This led him on to do some fairly simple sums and to come to the humungous realization that the Earth was a bit older than the official 4004 B.C. Creation date everybody accepted at the time. "Older" in fact, as in: "older than you could possibly imagine." Hutton went so far as to say that he couldn't imagine a beginning or an end, thus blowing everybody's geologic mind.

Or would have, had the book he wrote to explain all this *(Theory of the Earth,* 1795) not been so turgid and incomprehensible. Fortunately, after his death in 1797, his pal Playfair rewrote the whole thing to make it readable, and in doing so (a) made Hutton famous and (b) kicked off modern geology. It was reading the readable version that inspired Robert Jameson, another Scots rock hound, to take up his pen and attack Hutton's theory that rocks had started as hot magma coming to the surface from a molten interior. In 1804 Jameson graduated to the job of professor of natural history at Edinburgh, where his anti-Hutton lessons bored the visiting young Darwin so much, the lad later skipped all geology lectures when he went to Cambridge. Jameson spent many years defending the ideas of his former teacher, German Abraham Werner (Freiberg School of Mines) who had said that rocks had been deposited out of the waters of an Earth-covering ocean. Werner's work also excited American do-gooder William Maclure, who, after retiring rich at the age of thirty-four, devoted most of his life to traveling all over Europe, the Caribbean, and large tracts of America, picking up rocks—and, in 1809, producing the first geological map of the eastern United States. While building up to this, in 1805 Maclure was in the home of big rocks, Switzerland, when he met and was struck by the work of Heinrich Pestalozzi, who ran a primary school at Yverdon.

The Pestalozzian system (the basis for all modern primary education) was a radical departure: no books, lots of

but of the ideological will of the political state: citizen armies. You can see why Napoleon fell for the idea and used it to take over Europe. After publication, Guibert was invited to attend Prussian army maneuvers by another fan, Frederick the Great, who thought the *Essay* was one of the few pieces all generals should study. Frederick was the philosopher-prince in spades: He won all his battles; was a flute player and composer; kept his own orchestra; employed Voltaire as in-residence thinker; abolished torture; introduced compulsory education; instituted reformed legal processes, better communications, religious tolerance, and a free press; and beefed up the Prussian Academy of Science enough for it to attract some of the greatest minds in Europe. In 1772 the young Johann Bode was invited to work at the Academy's observatory in Berlin.

That same year, Bode publicized what became known as the Titus-Bode formula (Titus was the other guy), for the approximately geometric progression of distances between the seven planets known at the time. Taking Sun-to-Earth distance as 1.0, Sun-to-Mercury is 0.4, Sun-to-Venus 0.7, Sun-to-X 2.8, Sun-to-Jupiter 5.2, Sun-to-Saturn 10.0, and Sun-to-Y 19.0. "X" correctly predicted the position of the asteroids and "Y" that of Uranus, to be named by Bode after being found by William Herschel in 1781. William's sister Caroline had been intended by her mother as the family servant, because her stunted growth (four feet three inches) thanks to an attack of typhus meant she would never find a husband. In 1757 Caroline left Germany for England with her brother and became his lifetime apprentice astronomer, grinding lenses, keeping records of all his observations, compiling star catalogs, and then graduating to her own discoveries (several nebulae and eight comets). Caroline did so well that George III granted her a pension, the first in history for a female scientist. In 1835, and already age eighty-five, she finished compiling her nephew John's catalog of twenty-five hundred nebulae. In recognition of her work she was made one of the first female honorary members of the Royal Astronomical Society, together with Mary Somerville, the other amazing woman of the century. Self-taught mathematician, astronomer, botanist, mineralogist, geologist, and speaker of Greek, Mary ended up with the biggest address book in nineteenth-century science, including hotshots like Babbage, Humboldt, Davy, Arago, and Gay Lussac. In 1831 she translated Laplace's *Celestial Mechanics* and then went on to become the thinking man's fancy as a major science writer. Not for nothing is Somerville College, Oxford, so named. Her Laplace work really impressed Faraday, already the great electricity guru for his discovery that a magnet will generate current in a coil of wire.

Faraday ran the Royal Institution, and his 1860 Christmas lectures for children were transcribed by his assistant William Crookes.

exercise and music, children taught by children, no corporal punishment, no rote learning, and above all, show-and-tell. Children were given the chance to experience things before being taught how to read or write about them. It was essential that, in this way, the child learn to make judgments based on his or her personal experiences. Pestalozzi schools attracted visitors from all over the world. Early on, one devotee was Joseph Neef, former theology student turned soldier who'd left the army when a bullet stuck between his nose and right eye and he suffered nonstop headaches. By 1802 he was running a Pestalozzi school in Paris and giving tours to Napoleon and Talleyrand—and to William Maclure, who persuaded Neef to take the Pestalozzi gospel to the United States. Neef finally arrived there in 1806 and, after setting up three schools (two in Pennsylvania and one in Kentucky), in 1824 he was persuaded by Maclure to try the utopian community founded at New Harmony, Indiana, by English libertarian and textile tycoon Robert Owen. Things utopian came to the usual end, with internal arguments about who was in charge. In 1827 everybody went their own way.

Owen went back to a life of socialism in England, Maclure to retirement in Mexico, and Owen's eldest son, Robert Dale Owen, to yet another utopia, founded for freed black slaves at Nashoba, Tennessee, by English feminist Frances Wright. Robert Dale left his trace in two ways. Under the influence of Wright, he wrote and published the first-ever American pamphlet on birth control. Also, after three terms in the Indiana legislature, he got himself elected to Congress, where, as the democratic representative from Indiana, he fathered the bill to authorize the acceptance of the Englishman Smithson's bequest, used to found the Smithsonian Institution. Defeated for re-election after two terms in the House, in 1853 Owen was appointed American chargé d'affaires to the court of Naples, Italy. There, in 1855 a young fellow American turned up and was transfixed by Owen's political view of the world. Andrew White then returned to the States and became an academic, abolitionist, and New York senator. As chair of the committee on education, in 1864 White met and liked a self-made, ex-carpenter millionaire who was lobbying for the establishment of a new state university that would benefit from the Morrill Land Grant, would be nonsectarian, and would include mechanical arts on the curriculum. The former car-

TRACK ONE

Crookes would become a physics star in his own right, and a supporter of all things psychic (séances, antigravity, ethereal music, and a ghost called Katie King he fell for). At one point, Crookes did an investigation into the effects of cold on radium radiation together with James Dewar. Who was the king of chill, once he'd invented (in 1892) the vacuum flask to store liquid gas, and then expert on frigidity effects on phosphorescence, color, and the strength of materials. You name it, Dewar froze it. World War I stopped this expensive fun, so he turned to cheaper entertainment and began making bubbles, keeping one intact for three years and spending that time finding out about thin-film properties.

END TRACK ONE

penter offered half a million bucks. A deal was struck, and Andrew White became the first president of the new establishment, named, after the carpenter, Cornell University. Ezra Cornell (humble beginnings: farm, grocery store) had made his millions on the fringes of telegraphy.

Back in 1839, when Cornell got the Maine sales rights for a new plough, he'd met a guy who'd contracted to lay Samuel Morse's cable from Baltimore to Washington for the great 1844 Congress telegraph demonstration (when Morse transmitted the immortal one-liner: "What hath God wrought"). Cornell designed a trench digger to automate the cable laying, and then destroyed it to give Morse the excuse for late delivery. With this kind of foot in the door, Cornell then proceeded to lay, or string, telegraph wires all over the eastern states. The telegraph company shares he bought early on gradually rose in value from speculative to what you've always dreamed of. At the start, the telegraph wires were insulated with a mix of thread and shellac, but when talk began of laying a telegraph cable under the Atlantic, something better was needed. Gutta-percha would resist all seawater incursions. The growing demand for undersea cable soon made it essential to find a way to automate gutta-percha coating.

This was why in 1879 Matthew Gray produced his extruder, where hot gutta-percha was fed into one end of a container, in which a rotating screw delivered it to the other end, where under steady pressure it was pushed through specially shaped holes, to deposit a continuous layer on the cable running through a hole down the center of the extruder.

END TRACK TWO

AND FINALLY . . .

When plastics were discovered in the early twentieth century, Dewar's thin-film science and a version of Gray's extruder came together to produce the first food insulation material: clingwrap.

FOURTEEN

1790: *PHILADELPHIA GENERAL ADVERTISER* TO CHEMOTHERAPY

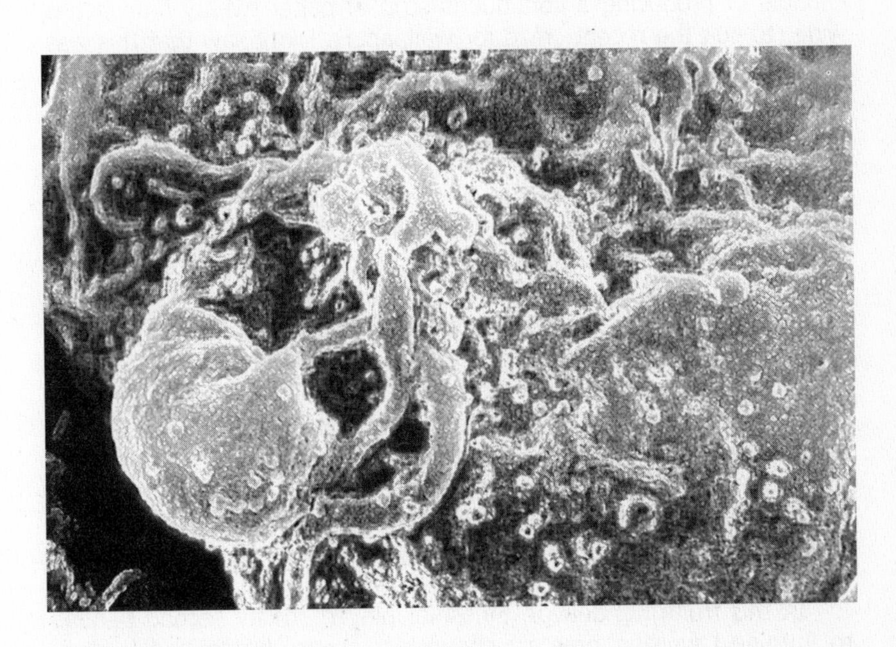

In 1790 the *Philadelphia General Advertiser* was founded, with the ambiguous masthead: "Truth, Decency, Utility." President Washington was in his first term, and the political parties were still finding their feet, looking for ways to get their message out. The *Advertiser* did so for the Republicans. It began by reporting congressional debates and general goings-on. However (one man's "Utility" not being the same as another's), the newspaper turned more and more partisan as time went by.

The paper's founder was Benjamin Franklin Bache, who spent a formative seven years with his grandfather (you know who) in France and Geneva. The effect was to leave him with a somewhat uncritical view of all things French and revolutionary. None of which got in the way of his tendency to publish any dirt he could find on American Federalists in general and Washington in particular. Bache might have picked up some of his revolutionary fervor back in Paris, in 1785, when his grandfather had apprenticed him to the Didot printing house. In 1796 Didot St. Leger had labor shortages because all the workers were in Napoleon's army. So Didot's mill manager Nicholas Robert came up with a design for the first-ever papermaking machine capable of producing a continuous strip of paper twenty-four inches wide (to suit the recent craze for wallpaper). Didot saw that this was really a machine for printing money and jumped at it. Several government grants and arguments about patent rights later, in 1802 Didot left Robert in charge of the mill and disappeared for seventeen years, in England with investors.

Robert's only other claim to fame had been in 1783, when he was onboard the first hydrogen balloon, built by his pal, poor Jules César Charles. He was known primarily as cuckold *du jour* because of his wife's very public affair with the poet and statesman Alphonse de Lamartine. Even as Charles ballooned along, Muesnier de la Place, engineer, soldier, and water-desalinization expert, realizing this balloon thing was going to turn out to be a military space race (the first crews up there could shoot down on others below), was designing a dirigible, powered by hand-operated propellers (the idea never got off the ground). Muesnier gave national security a boost by making a valuable contribution to the research effort into how to make lots of hydrogen before anybody else discovered the same (especially the Brits).

In this fruitful endeavor, Muesnier played strictly second banana to Antoine Lavoisier, greatest chemist anywhere by anybody's standards (he set up modern chemical nomenclature, named oxygen, and identified oxygen as essential to combustion). Being a public-spirited chap (just as well, now that the revolutionary junta were running things), Lavoisier also headed the gunpowder-manufacture department for the government, and he sat on committees dealing with everything from the location of slaughterhouses to the adulteration of cider, animal magnetism, and general industrial practices. And, of course, how to make cheap hydrogen (he decomposed water, and voilà, there it was). Alas, in 1794 all this do-gooder stuff wasn't enough to save Lavoisier's head from the guillotine.

The *Advertiser* was pro-France (the recent friend, bankrolling the War of Independence) and anti-Britain (the recent enemy). These leanings were soon to be evident, when, in 1783, the new French minister to the United States arrived to a rapturous *Advertiser* welcome. Minister Genêt's welcome from the U.S. government was a little more muted. The recent execution of the French king by zealous revolutionaries had left President Washington feeling less than comfortable, and besides, relations with Britain were now on the mend. And now here came Genêt with plans for a Franco-American economic alliance that would very aggressively sock it to the Brits.

Possibly in an attempt to show he really meant business, not long after he settled in, Genêt (at his own expense) fitted out a captured British ship (the *Little Sarah,* renamed as the *Little Democrat)* and sent it out to attack British shipping. Genêt did so without official American sanction, from the port of Philadelphia. When George Washington heard about this bit of freelance muscle flexing, he went ballistic and promptly asked the French authorities for Genêt's recall. However, given that things in Paris had changed even in the short time since Genêt had left (i.e., different, and even less-democratic, kinds of revolutionaries were now in charge there), it became rapidly clear that the Minister's "recall" was going to lead to the minister's "death by guillotine." So Genêt was allowed to stay and promptly secured his position by marrying the daughter of George Clinton.

By this time (1794), Clinton was on his fifth (out of six) New York governorships and a whisker away from the presidency. Alas, whisker is how it stayed, though he got the vice presidency (twice). Clinton was the politico's politico, using every trick in the smoke-filled back room, from patronage to states' rights to "interests of the common people," so as to stay in power and direct the affairs of New York. This he did for over twenty years, steering the place from colony to state with hardly a ripple. Clinton had started out as a militiaman, and in 1760 he was at the siege of Montreal, where the British general Jeffrey Amherst finally finished off French Canada and with it a long-cherished Gallic dream of uniting Canada with Louisiana.

The Montreal event was (in the event) something of an anticlimax. Amherst coordinated the arrival of three armies

In spite of this, we know what his face looked like, thanks to his portrait (with his wife) by Jean-Louis David, the great neoclassical painter. David had been a great rococo painter till he went to Rome and (as he said) felt as though cataracts had been removed from his eyes. David did things on a grand scale, in Greek or Roman mode, including various gigantic paintings involving (or for) Napoleon (who said: "For anything to be good it has to be big").

In these terms, David's best effort was probably the monumental *Coronation of Napoleon.* The canvas included hundreds of figures, including seventy VIPs the viewer (and they) could recognize, some of whom weren't even present at the time (like Napoleon's mother, who didn't like Josephine). If David had a weak suit, it was getting proportions right, so he hired Ignazio Degotti, an Italian stage designer working at the Paris Opera, to mark out the background perspective. The completed painting was so true to life it was the nearest thing the period could offer to a group photograph.

Ironically, one of Degotti's students was none other than Daguerre, who would make his name with the group photograph. After twenty-two years as a stage designer, Daguerre opened the Paris Diorama (a rotating auditorium moved the seated audience slowly round, bringing them in front of different scenes: the Alps, a city street, and the countryside, complete with special effects such as fog, sunshine, and night-lighting effects). Bored with this (and anyway, the place burned down), in 1835 Daguerre found a way to make images chemically (light-sensitive silver iodide on a copperplate reacted to an image shone on it, and the image was then fixed with a salt solution). It worked brilliantly well, as long as the sitter didn't move a muscle. Named the daguerreotype, the result was a sensation.

It certainly blew away Samuel Morse, who was in Europe at the time and who took one of Daguerre's cameras back to New York, set up a studio, and spread the word. He was soon able to do this in really spectacular fashion with his own invention. Once, that is, he'd managed to put together Joseph Henry's electrical know-how (unattributed) with Alfred Vail's code (unattributed), and use Ezra Cornell's plough (attributed) to bury a telegraph line and then (when the insulation failed) string it on poles all the way from Baltimore to Washington, so as to stun members of Congress on May 24, 1844, with his famous instantaneous-transmission dot-dash message. Cornell, a successful carpenter and grocer, saw what was coming (or rather, where the railroads were going), landed the contracts for telegraph poles that would follow the railroads, then bought shares in what became Western Union, and ended up (what else) so rich that in 1868 he was able single-handedly to endow his own eponymous university.

totalling seventeen thousand men, so far outnumbering the French defenders that they put their guns down without a fight. Many of Amherst's troops had sailed up the St. Lawrence after the fall of Quebec, a few months earlier. And they'd managed this, in spite of the fact that the French had removed all the marker buoys, thanks to the good work of two surveyors, a Swiss and a Dutchman, both members of the Royal American Regiment and both hot stuff with mapping and measuring. Samuel Hollandt and Joseph Desbarres surveyed the river ahead of the troopships and were so highly regarded that, as a result, they were commissioned to do a giant survey of the entire American East Coast (you never knew, one day it might be useful for landing troops, in the wildly unlikely event of a war with the colonies). Their finished masterpiece was titled *Atlantic Neptune* and was a magnificent work, beautifully engraved and illustrating every nook and inlet along the Eastern seaboard from Nova Scotia to the Gulf of Mexico. Alas, the material didn't make it out of the printer's hands until 1777, one year after that wildly unlikely event.

While Des Barres and Hollandt were measuring various bits of Canada, prior to the attack in Quebec, they were joined by a young navy type who seemed a natural with a theodolite, by the name of James Cook. Cook then took his newfound skills to the Pacific in 1768 (with a shipload of observers) to measure the transit of Venus across the sun. He then set off to look for the Great Southern Continent, which was thought to exist because such a land mass would be required to balance the continents in the Northern Hemisphere. Whatever it was, Cook didn't find it. After a year of charting, he returned to the United Kingdom, filled up two more noodler boatloads, and went back on another search for the G.S.C. ("Not there," he said, on his return.)

Cook's first trip had included Daniel Solander, a naturalist (he paid his own way) who was in England to spread the word about fellow Swede Carl von Linne's (a.k.a. Linnaeus's) classification system (you know: *Hedera quinquefolia* and all that). Solander did such a good job that he was given one at the new British Museum. And he never went home, especially when he discovered he'd been discovered by London society, avid for tales of exploratory derring-do (he'd slept in the Tierra del Fuego snow one night and nearly died), and was the toast of all fashionable dinner parties. One of these was given for him by Elizabeth Mon-

One of the first railroads followed by a telegraph line was the New York and Erie. In 1851, the railroad superintendent, Charles Minot, was stuck at Harriman, New York, waiting for the eastbound train to pass so that his train to Goshen could continue (the problem was that with only a single track you either waited or crashed head on). Minot had the idea of telegraphing Goshen to see if the other train had already left. It hadn't. "Hold the train there," he sent, and drove his train on down the track himself (the engineer didn't believe this black magic and refused), thus inventing railway signaling and everything it made possible. Back in 1846, the Erie Railroad track had been laid in the first place thanks to a risky contract to supply twelve thousand tons of rail at half the going (import-from-Europe) price, signed by an iron-and-nail foundry close to bankruptcy and desperate to stay in business. The Lackawanna Iron and Coal Company of Slocum Hollow, Pennsylvania, was surrounded by deposits of anthracite fuel and iron raw material. Once they had the rail-making contract, all they needed was a way to ship out the finished rails (the first mass-produced rails in the United States).

In 1851 the Delaware, Lackawanna, and Western Railroad was formed to do so, and the company (and its owners, the Scrantons) went on to fame, fortune, and a posh boardroom. The boardroom lacked only a boardroom painting depicting something about the company and its effect on the landscape. In 1855 the company commissioned a picture (for all of seventy-five dollars) from a struggling young nobody, George Inness. George was not long back from a spell in Rome and would eventually become one of America's great painters. His *The Lackawanna Valley* (with its view of the trainload of rails coming up the valley) became an influential American art classic.

One of the many influenced was artist Louis Eilshemius, whose career was somewhat less smooth. Eilshemius did all the right things: Cornell, Art Students League in New York, art academy in Paris, private lessons in Antwerp. This was unfortunately followed by a life of increasing mental problems, starting around 1911 when he gave up the academic style and started painting on newspapers, cigar-box lids, etc. Eilshemius talked loudly to his paintings, specialized in tortured nudes (ruffling feminist feathers), and wrote articles for the *New York Sun* on everything. By the time his work was being truly appreciated, he was truly nuts. His older sister was altogether unlike him. And, unlike Louis, she changed the world.

Fanny Angelina Eilshemius went to Swiss finishing school (her mother was Swiss) to learn French and home economics. On a later visit to Europe with her parents, Lina met and then, in 1874, married Walther Hesse, a district medical officer working in the mountain mining communities on the German-Czech border, with an interest in air

tagu—rich, witty, loaded with jewels, and not looking her age. If you got invited to her Mayfair house, you were a trendsetter—as were the others round the table that night in 1775, including Dr. Johnson (the doyen of letters) and Sir Joshua Reynolds (the doyen of the paintbrush). Reynolds, now at the height of his career as England's greatest portraitist, was seven years into his twenty-year role as president of the new Royal Academy and dictator of the art world. Reynolds's secret was to put his sitters in some evocative setting, which worked to the tune of 150 sitters a year. If you mattered, Reynolds painted you: royalty, aristocracy, politicians, artists (and Cook and Amherst). In 1782 he painted twenty-two-year-old William Beckford, who'd just written an oriental Gothic novel, *Vathek,* as full of weird imaginings as was Beckford.

Left an unimaginable fortune, the boy was busy spending it in the most profligate manner possible: traveling the continent for twenty years with an entourage, collecting anything that took his fancy, especially books (and possibly boys), most of which would eventually stack up in the 1794 pseudo-Gothic extravaganza he constructed. Fonthill Abbey (complete with medieval tower, mullioned windows, and the whole Perpendicular bit) was intended to replace the perfectly acceptable Palladian family home Beckford had torn down. He surrounded the Abbey with a high wall, behind which he then shut himself for twenty years with a doctor, majordomo, French priest, a zillion books, and assorted bric-a-brac (and possibly boys). At one point, back in 1792 in Lausanne, while taking a break from travel to read his way through a library he'd just bought, Beckford threw a spectacular shindig on the lakeside. He commissioned a French architect and landscape designer, Joseph Ramée (who was on the run from the revolutionaries back home), to do the party decorations. Ramée was a wanderer, designing and building as he went through Belgium, Sweden, Switzerland, Germany, Denmark, and (for four years from 1812) the United States. There, he left an indelible mark with the wonderful Union College, Schenectady, the first unified-plan campus in the country and probably an influence on Jefferson's later ideas for the University of Virginia. Less well known (and why not) is Ramée's lumpy, Doric-temple mausoleum at Ludwigslust, Germany, created to house the remains of the princess of Mecklenburg-Schwerin, daughter-in-law of the grand duke of the same

TRACK ONE

quality (many of his miners had lung diseases). In the hot summer of 1881, he was doing studies on bacterial air contamination by sucking air through a glass tube coated with gelatin, which trapped the bacteria and made the development of cultures possible. Or rather, it did so when the gelatin wasn't melting in the heat. Lina suggested Walther try an ingredient she used as a gelling agent for soups, blancmanges, jellies, and stuff. It was a seaweed derivative, agar-agar, and it turned out to be the ideal nonmelting, nontoxic substrate for growing bacterial culture. Walther told his boss, bacteriologist Robert Koch, who was searching for the origin of tuberculosis at the time. Koch used the agar, found the tubercle bacillus, and got the glory.

END TRACK ONE

place—who, like many of the same rank, had his own orchestra.

It included first violinist Leopold August Abel, who had been taught by Bach. Leopold had two sons, one of whom was a violinist. The other became the duke's miniature-portrait painter. The painter's son, unimpressed by local prospects, lit out for greener pastures in England. There, in turn, *his* son Fred (is this getting a bit too genealogical?) revealed a taste for risky experiments. After qualifying himself in the new Royal College of Chemistry, Fred became involved with munitions and was soon the British War Department's expert on bangs and what happens, chemically, during them. In the course of this research, he succeeded in putting the Brits ahead in the race to produce big explosions by coinventing cordite with vacuum-flask guru James Dewar.

The man who had set Fred on the road to this military mayhem was Professor August Hoffmann, since 1845 the first director of the college where Fred had studied and the instigator of the great what-colors-can-you-make-from-coal-tar craze that fired all his fellow chemists, once the gaslight manufacturers started chucking away the coal tar (the by-product of gaslight making) into any nearby repository. One of the earliest of Hoffmann's new grads to fiddle with the stuff found the first aniline dye ("Perkin mauve," with which the first postage stamp was colored), and others followed suit. This included Hoffmann himself, who produced the highly fashionable "Hoffmann violets." In 1870 the director of the BASF research lab near Frankfurt (Heinrich Caro, who had worked in England until he saw that the chemical future was going to be back home in Germany) studied how Hoffmann had done what he'd done, and then proceeded to discover his own personal coal-tar derivative dye: methylene blue.

END TRACK TWO

AND FINALLY . . .

Some time around 1882, Paul Ehrlich, a German medical researcher working with Koch and using Fanny Angelina Eilshemius's agar tissue cultures, discovered that methylene blue would stain bugs. It was this stain that revealed the tubercle bacillus in all its glory. Intrigued that dye didn't stain some cultures, Koch guessed that they might be absorbing the dye, which meant you might be able to do things to the bugs with the dyes. After seven years and 605 different dye prepararions, Ehrlich came up with "Preparation 606" (a.k.a. Salvarsan), the first "magic bullet" treatment for disease (in this case: syphilis). In 1908 Ehrlich got the Nobel for inventing chemotherapy.

FIFTEEN

1664: LENS GRINDER TO HAIRDRESSING

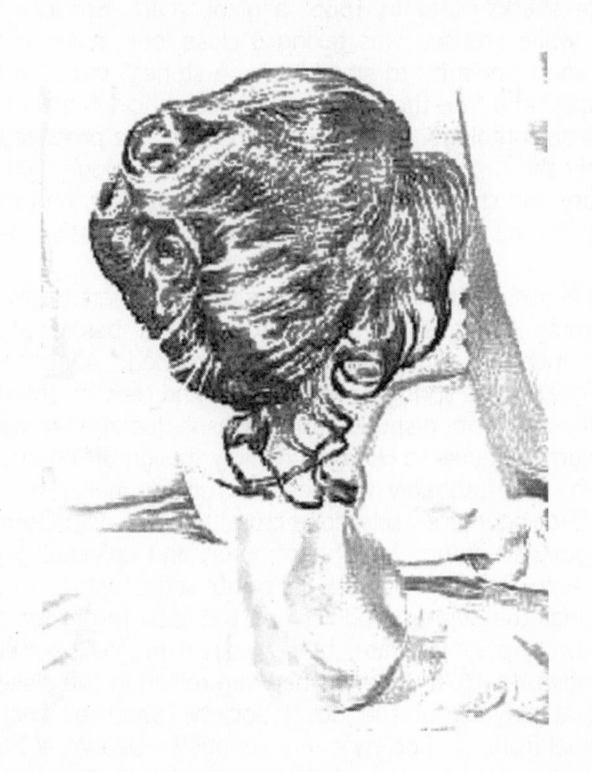

Long-focus lenses are the way to look far and see much. So in 1664, when an Italian peasant-turned-instrument maker invented a lens-grinding lathe that allowed him to create lenses with a focal length up to 150 feet, all of a sudden you could use humungously-long telescopes and see very distant things, like Saturn's moons and more. Thank you, Giovanni Campani.

Campani became the instant favorite of Ferdinand II, grand duke of Tuscany, who was a science buff because he'd been taught by Galileo. Ferdinand and his brother Leopold funded some major minds to investigate the latest hot topics (the vacuum and astronomy) in a kind of club of nerds named the *Accademia del Cimento* (motto: "double-check everything"). One of them was a visiting Dane named Niels Stensen, who turned up in 1666 in search of work and was welcomed with open arms because he had already discovered a major salivary gland (Stensen's duct). As you might expect, he now started to investigate sharks. Late in 1666, a giant shark was caught off Livorno, and while Stensen was taking a close look at its teeth, he noticed a strange similarity to small "tongue stones" found in Malta. Such is science. "Aha," he thought, "not stones—ancient shark teeth." And invented paleontology. Eighteen months later, he produced what was effectively the first book on the principles of geology, with sedimentary theory and stratification sequences and all that. And then he gave it all up for a vow of poverty, then died of starvation. Such is life.

At least he missed the plagiarism fuss, when Robert Hooke went ballistic, claiming to have discovered fossils well before Stensen. Hooke spent much of his time yelling: "I wuz robbed." About Stensen (fossils), Huyghens (the spring-driven watch), and Newton (the law of gravity). All these priority disputes, about was-it-Hooke-or-somebody-else, only occurred thanks to Hooke's inability to sign off on anything. Nonetheless he was probably the best instrument maker the Royal Society ever had: odometer, telescope crosshairs, iris diaphragm, spiral gear, air gun, air pump, wheel barometer, and universal joint, to name but a few. He also worked on combustion, optics, and thin films. Hooke was the supreme fiddler and the ideal target for satirist Thomas Shadwell, poet laureate, better known for his beer-swilling and opium habits. In 1676 Shadwell had 'em rolling in the aisles with *The Virtuoso,* a play about the Royal Society (Shadwell hilariously rewrote material from the Society's very serious *Philosophical Transactions*), starring two weirdos remarkably like Hooke: Sir Nicholas Gimcrack and his sidekick, Sir Formal Trifle. The play's "virtuoso" (a.k.a. scientist) was a guy who "broke his brains about the nature of maggots" and did meaningless experiments like weighing air and taking swimming lessons on a table. The same year, Shadwell also brought out a tragedy, *The Libertine,* for which the incidental music (including the famous "Nymphs and Shepherds") was written by a young genius, Henry Purcell, who was only eighteen and only lasted till thirty-six. During that time, he packed a lot in and became one of

In January 1665 Campani's lenses were described in the inaugural edition of the French *Journal des Savants,* the first-ever science journal, edited by Denys de Sallo, who thought it was time to keep people abreast of the latest discoveries. Sallo started off as a lawyer, then went into politics. He read voraciously and employed two full-time secretaries to whom he dictated his thoughts. The result was many publications on subjects ranging from science to how to address the queen. His *Journal* only lasted for thirteen issues before being suppressed for questioning authority. In 1702 the *Journal* was reintroduced (after having been checked out by the government to make sure nothing was included that might offend the king or the church) under a new editor, Jean-Paul Bignon. Bignon was a well-connected cleric with (politically correct) science leanings, and was also president and vice president of the French Academy of Science for over forty years, the king's librarian (he opened the library to the public for a generous three hours a week), dictator of all French research, and author of *The Adventures of Abdallah.* So to put it mildly, you stayed on the right side of Bignon. This may be why, in 1710, Bignon received a present of a pair of amazing spider-silk gloves from one of his protégés, René-Antoine de Reaumur, one of the new technocrats dedicated to putting science at the service of the state (and Bignon).

Reaumur was then given the editorship of a large industrial encyclopedia and rapidly became an expert on the production of iron, steel, tinplate, and porcelain (he invented a type of ceramic that is still used on rocket nose cones). Reaumur also became the bee's knees on bees and wrote volumes on insects (see spider-silk gloves) and pest control. Reaumur also got involved in the vexed topic of spontaneous generation (for instance, how did worms regenerate themselves after being cut in half?). As part of this work, he investigated the parthenogenetic tendencies of aphids and suggested that the only way to confirm that they reproduced asexually was to rear them in isolation from birth and watch to see if they had babies. This suggestion was taken up by a young Swiss, Charles Bonnet, who in 1746 reared a female spindle-tree aphid all by herself and observed she had ninety-five offspring. All by herself. The news shook the scientific community. As did Bonnet's next

England's greatest composers, turning out more than fifty pieces for the stage, as well as work ranging from songs, anthems, and organ works to funeral odes and the first nonspoken masque: *Dido and Aeneas* (from which, try "Dido's Lament": unforgettable). Purcell has been described as the Shakespeare of English music and not without reason.

In 1679 Purcell became organist to the Chapel Royal and two years later met Bernard Smith (Bernhardt Schmidt), who was busy repairing one of the king's organs and would end up building great organs all over England. Purcell then found himself embroiled in the Great Organ Battle. In 1683 Smith and his archrival Renatus Harris were asked to compete for the building of an organ for the lawyers at the Temple (very well-heeled crowd). In the end, it came down to a competition, in which Purcell played Smith's instrument. Nobody could make up their minds whose organ was better, so it was left to an impartial judge, who chose for Smith. A hung jury was a new experience for the judge in question, George Jeffreys, that year Lord Chief Justice of the King's Bench. Jeffreys was a lout, known for his tendency to be drunk in court and to be coarse and bullying during cross-examination, and a sadist who liked to offer his prisoners leniency in return for a guilty plea, and then hang them—often, on the spot. His execution total was estimated at six hundred, which earned him the name "Hanging Judge Jeffreys." A real sweetheart. Fortunately he got his comeuppance shortly after helping James II with a purge on Protestant clergy that was unpopular enough to force the king into exile. Jeffreys was on his way out of the country, too, when he popped into a pub for a last drink, got caught, and ended in the Tower of London. One of the purged priests was antipapist Henry Compton, bishop of London, who was reinstated and crowned the new monarchs (William and Mary).

Compton was also a botany freak and grew all kinds of rare plants in his gardens at Fulham Bishop's Palace (the gardens are still there). The rarest plants came from America, where his seed supplier was Alexander Spottiswood, lieutenant governor of Virginia. Spottiswood had been in Georgia since 1710 and was rapidly amassing good things (like eighty-five thousand acres of property and the biggest ironworks in the British Empire). He was also deputy postmaster to the colonies, hired Ben Franklin, and wiped out pirate Captain Blackbeard. Spottiswood's seed-collecting and dried-specimen legwork was done for him by botanist and ornithologist Mark Catesby, who visited the colony twice, on commission from growers and collectors in England. Catesby rooted around in Appalachia, the Bahamas, and Jamaica, and ended up with enough data and drawings to put together two volumes which became the last word on *flora Ameri-*

trick: cutting a rainwater worm into twenty-six pieces and ending up with twenty-five new worms. Bonnet was chief exponent of the "Great Chain of Being" theory that all organisms had been created at Creation in a universal, graduated series from slime (bottom) to angels (top). He also got close to photosynthesis with his work on plant leaves. His attempt at a scientific explanation of the soul was less successful. In 1777 Bonnet had a visit from another one of those English aristocrats on a European culture tour: William Beckford, who since the age of twelve (left a colossal fortune by a Caribbean plantation-owning father) had been the richest man in England. In 1796 he built himself a gigantic pseudo-Gothic abbey at Fonthill and lived there for twenty years with twenty thousand books, works by Rembrandt, Velázquez, and Titian, plus objets d'art of ever kind. In 1822, finally short of cash, he sold up and moved to a tower in Bath.

Back in the days before he'd settled down behind Fonthill's walls, Beckford was in Spain and Portugal, and met Luigi Boccherini, cellist extraordinaire, inventor of the string quartet, and at that time composer to the duchess of Osuna in Madrid. This, after shopping his talents around Lucca, Venice, Trieste, Vienna, Paris, Rome, Nice, Florence, and Modena. Going where the money was. A musician's life. In 1786, when Beckford met him, he'd just landed the kind of job he must have waited for all his life: writing stuff for Friedrich Wilhelm II of Prussia but without having to leave Spain. Friedrich Wilhelm was an amateur cellist, and his other tame composers were Haydn, Beethoven, and Mozart (whose Prussian quartets have a cello part difficult enough to suggest the king was no mean instrumentalist). Friedrich inherited a considerable musical talent from his uncle Frederick the Great—and little else. Most people thought him an easygoing, self-indulgent wimp. It was perhaps just as well he fell off his perch in 1797, before Napoleon could test him with invasion and war. Alas, he was only a few days from the end when the new American minister arrived. So John Quincy Adams had to kick his heels till Friedrich Wilhelm III took over. Adams threw himself into German literature and philosophy (his ambition at the time was to become a writer) and in 1799 started a night-and-day effort to translate a novel about the fairies: *Oberon,* by Christoph Martin Wieland. By 1801 Quincy was putting the final touches to the manuscript when he heard that another version was

cana. When the great categorizer Linnaeus went to Oxford in 1736, one of his stops was at Catesby's collection. Linnaeus was a Swede with one thing in mind: making order out of the chaos of natural history caused by the avalanche of new species from explorations East and West. So in 1753 he produced a giant list of eight thousand species, all neatly named according to species and variety, as in: *Rosa* (rose) *odorata* (fragrant). He amassed a gigantic collection in Uppsala, Sweden, where he was professor, and then died. After several attempts to sell the collection, his trustees off-loaded it onto a grateful James Edward Smith of Norwich, England, who translated Linnaeus's books and in 1788 founded the Linnaean Society. That same year, he gave a paper on "The Irritability of Vegetables" and went on a nine-year tour of European botanical gardens and museums to meet other botanical geeks. One of whom was Paolo Mascagni, who would become famous for the first life-size, show-and-tell anatomy kit: an assembly of three giant colored prints, showing muscles and blood vessels in total cutaway detail.

On Mascagni's death in 1815, the family asked his assistant, Francesco Antommarchi, to get the kit published. In 1819 Antommarchi was hired as doctor to Napoleon (in exile on St. Helena), sailed away with the pictures, and later produced his own plagiarized versions. On St. Helena, Antommarchi replaced the disgraced Barry O'Meara, who'd been there since Napoleon's arrival in 1815 and had been given a pay raise in return for extra duties: spying on and reporting everything Napoleon said. After the island's Governor Lowe complained about his work, O'Meara withheld these secret reports and was fired, returned to the United Kingdom, and in 1822 (after Napoleon's death a year earlier, in suspicious circumstances) caused a storm with a book about Lowe's mistreatment of the emperor.

O'Meara then took up with radical politician Daniel O'Connell, thorn in the side of the British government, in and out of Parliament. From 1821 until the end of his life, this Irish lawyer-turned-nationalist fought for Irish independence and above all to improve the condition of the starving Irish peasants, living on rented land, suffering absentee landlords and their gouging agents, and subsisting on potatoes. For twenty years, O'Connell warned of the dangers of this single-crop dependence, and in 1845, with the arrival of the potato famine, was vindicated. Originating in ships from the United States to England, spores were blown across to Ireland and soon destroyed the crop—and went on doing so for six years. At the end of this, a million were dead and another million emigrated. Arguments raged about what had caused the blight (fungus? insect? human?) and what to do about it.

In France a special commission reported in favor of the fungus

just out, and that Wieland was crazy about it. So he turned his attention back to statesmanship and, eventually, to becoming sixth president of the United States. His Wieland translation only made it to the bookshelves a hundred years later (don't rush).

The urbane and slightly overpolished Wieland may not exactly be your cup of tea, although there are those who think of him as the Voltaire of Germany. After a few years experimenting with blank-verse tragedy, writing lightly erotic verse, and developing a tendency to get nearly married rather frequently, in 1772 one of his novels had come to the attention of the dowager duchess of Weimar, who invited him to come teach her son, Karl August. Wieland accepted, got married and had fourteen kids, and settled into sleepy Weimar. It wasn't sleepy for long, however, once Karl August became duke and was determined to put his little spot on the map. First thing he did (good move) was hire the great philosopher, writer, and general genius Goethe to run the place better. Goethe's improvements included cutting the army from 520 to 142, raising intellectual standards at the Weimar duchy's university in nearby Jena, and kick starting local industry (textiles, glass, and mining). By 1800 the Duchy was the literary center of Europe and cradle of Romanticism. Things went less excitingly on the industrial front, in spite of the tireless efforts of Johann Wolfgang Dobereiner, a chemist imported by Goethe to beef up the technical side of Jena University. While Dobereiner was on the faculty, he also taught chemistry to a few artisans and technicians. He worked on processes for turning potatoes to sugar and conducted a series of gaslight experiments, which only ended in 1816 with a giant explosion. Dobereiner also investigated the table of elements, invented a gas lighter, made improvements to wine making, and received the Cross of the White Falcon for all these efforts. One of these included teaching former apothecary Frederich Runge, who went on to change the history of medicine and possibly save your life.

Runge made his mark, but not with his work on belladonna or his discovery of caffeine, nor even with his suggestion that creosote would save a zillion trees (in America, above all, it did so) because it would increase the working life of a wooden railroad cross tie by fifteen times. What really put Runge in every hospital's good books was his distillation of the gunk by-product of gaslight manufacture,

TRACK ONE

theory, whose proponent was Alselm Payen, by this time a science "big vegetable" (Fr. = "important person") whose first book, back in 1826, had been dedicated to the potato. Payen had gone on to become the world's expert on starch, then turned to investigation of plant tissue and then discovered the starchlike material filling plant cells. In 1839 the name for this—"cellulose"—was coined. In 1881, in Kew Botanic Gardens, London, Charles Cross and E. J. Bevan developed a way to process cellulose and turn it into a syrupy, viscous liquid, which if sprayed through nozzles would harden into the artificial fibers Cross and Bevan named viscose. In 1908 a young Swiss, Jacques Brandenburger, working in the clothing business, turned the stuff into a transparent film he called cellophane.

END TRACK ONE

coal tar. From which muck he extracted a substance that he named carbolic acid, whose antiseptic properties saved more lives than you could shake a scalpel at, thanks above all to Joseph Lister. Lister turned surgery from cross-your-fingers butchery to better odds, after he heard that carbolic thrown on sewage seemed to have killed the bugs infecting nearby cattle. Lister tried it mixed with water. Between 1865 and 1867 in Glasgow Infirmary, out of eleven compound-fracture patients (standard recovery rate, nil), no fewer than nine survived, thanks to antisepsis. In 1873, one of Lister's patients was a young poet who'd lost one foot to tubercular arthritis and was about to lose either the other or his life. Lister took him in for eighteen months, operated antiseptically, and saved both leg and life.

W. E. Henley wrote up this medical ordeal in *Hospital Verses,* and survived to become an important magazine editor and literary harrumph who would introduce and promote the work of H. G. Wells, Kipling, Yeats, Rodin, and Whistler—and, of course, that of the guy who had become his bosom pal after he visited Henley in hospital: Robert Louis Stevenson, who was about to give up the practice of law for writing. In 1876 Stevenson was in France and fell for an older American woman, Fanny Osbourne. In 1888 he and Osbourne were in San Francisco fitting out a yacht for life in the South Seas. It was during their sojourn in the Bay Area that that Stevenson wrote about the infamous head of the San Francisco Committee of Vigilantes, William Tell Coleman, ex-millionaire, ex-nearly presidential candidate, and one tough hombre—and, in 1882, in Death Valley, discoverer of colemanite, in major borax deposits.

END TRACK TWO

AND FINALLY . . .

In 1915 Charles Nestlé (real name Karl Nessler) wound women's hair in absorbent paper soaked with borax paste, held the paper in place with a cellophane curling tube, then used heaters to set the curl. In 1927, after DuPont improved things by making cellophane waterproof, Nestlé opened a permanent-wave salon in New York. Hairdressing never looked back.

SIXTEEN

1773: BOSTON TEA PARTY TO CONTACT LENSES

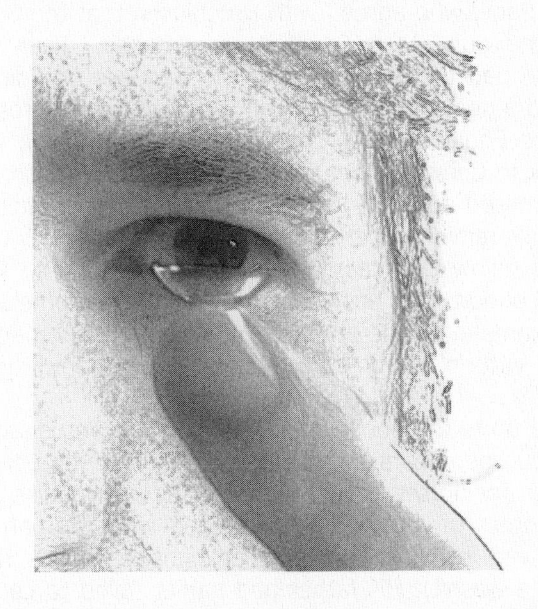

At around 7 P.M. on December 16, 1773, a group of men, variously disguised, boarded an English merchant ship, the *Dartmouth*, and in a quiet and orderly fashion, threw into Boston harbor waters a small fortune in tea (342 chests). The tea had been due for sale, at a hefty discount, direct to the customer, thus bypassing the local middlemen. Even including the customs duties to be paid to the Brits, the tea would have sold for much less than the smuggled variety usually drunk by Bostonians. So the local merchants were understandably irate. But the Boston Tea Party stirred up a lot more than wet tea leaves.

The tea was addressed to friends and relatives of Massachusetts governor Tom Hutchinson, which put him between a rock and a hard place. Hutchinson's handling of the affair got unfair press, considering he had always been a defender of local interests, proposed a plan of union for the colonies, and was critical of the arrogant way the Brits treated their American cousins. But Bostonians were too ticked off to put up with Hutchinson's judicious defense of the rule of law and of reasoned and gradual change. So when he was "recalled for consultation" to London in 1774, good riddance. Back in London, a year later at a musical soirée, Hutchinson met and became pals with a foreigner named Pascal Paoli, who agreed with Hutchinson that the Bostonians had jumped the gun. As a Corsican revolutionary, Paoli knew all about guns. He had begun by fighting the Genoese occupiers of his island, then did a deal with the French, who then double-crossed him. He then fled to England (where he first met Hutchinson), was then welcomed back to Corsica by the now-postrevolution French, and was then double-crossed again by the French (and in retaliation burned down Napoleon's family property). Paoli then invited the Brits to take over. Then his fellow Corsicans rose against the Brits, so Paoli was once again in London. This unswerving loyalty to the homeland acted as a shining example for one of Italy's more romantic Romantics, and in 1788 Count Vittorio Alfieri dedicated one of his many tragedies to Paoli.

Alfieri lived up to the poetic image: flamboyant and dissolute, living all over Europe, years of closeted study, attempted suicides, duels, a passion for horses, and a lifelong affair with Louise Stolberg, the tragic countess of Albany and estranged wife of Bonnie Prince Charlie (who lived in Florence and called himself Charles III of England, which he wasn't). His father had earlier failed to capture the English throne but still called himself James III of England (which he wasn't). Charlie's brother, Henry IX of England (which he wasn't) provided Alfieri and Stolberg with a silk-lined Roman love nest in one of his palaces. Henry was a well-meaning prince of the church, who sold the family jewels to help the pope pay Napoleon's tribute, and thus in 1799 was down to his last million—which the French then grabbed, so he hightailed it to Venice, where he sold the silver to stay afloat until a pension turned up from George III of England. This, thanks to importuning by Sir John Coxe Hippisley, diplomat in Italy at the time and who was later to become an M.P. and, in 1823, the star of the Great Treadmill Debate. Hippisley was a do-gooder who believed (extreme radical stuff back then) that prisoners should not suffer the backbreaking pain of the treadmill (to grind flour or drive looms),

TRACK TWO

The Tea Party was a disaster waiting to happen once the British Chancellor of the Exchequer decided to make up a budgetary shortfall by taxing American imports of tea. Then Parliament made revolution unavoidable by giving the American tea monopoly to the East India Company. The Chancellor was Charles Townshend, an aristocrat who had a fabulously rich wife and was part of the old-boy network and an amateur economist. Back in 1763, looking for a tutor for his stepson, he approached the Glasgow professor of moral philosophy, Adam Smith, who four years earlier had impressed Townshend mightily with his *Theory of Moral Sentiments,* in which Smith had explained how an "invisible hand" mitigated the individual's raw self-interest enough to work for the common good in an open marketplace. This laid the groundwork for his later magnum opus (in 1776): *The Wealth of Nations,* in which Smith invented the division of labor. While tutoring Townshend's stepson, Smith spent time in France and met the intellectual cream of the Enlightenment, who spent most of their talk time in salons hosted by various beautiful women, one of whom, Sophie de Grouchy, in 1798, would translate Smith's *Moral Sentiments* into French.

De Grouchy was vivacious and beautiful and (her enemies said) ruthless and ambitious. The salon she ran was internationally famous, and the philosophical chitchat was as often American as it was English or French. As things moved revolutionward, de Grouchy's place increasingly became a center for the far left and then an activist hotbed. All this (and de Grouchy's ambition) was to drive her mild-mannered mathematical husband into shark-infested political waters. The Marquis de Condorcet had started quietly enough, with an impressive paper on calculus written when he was twenty-one. His big thing was what he referred to as "social mathematics," the application of which would make it possible to predict social behavior, for the good of all. For two years, when his pal Turgot was French treasury secretary, Condorcet did good work for canals, cattle plague, social welfare, weights and measures, and a plan for state education that is still in operation today. If only his politics had been as good as his calculus. During the French Revolution, he backed the wrong splinter group and, in 1794, ended up dead in a cell. This was a marginally better end

which was dangerous, physically damaging, and, Hippisley believed, should be replaced by hand-cranked mills.

This view was backed by medical evidence from his colleague Dr. John Mason Good, poet, druggist, surgeon, and prison reformer, who spoke more than nine languages and also wrote on subjects as diverse as head-bump reading, philology, and Chinese grammar. In 1804 Good had started work on an encyclopedia with a self-taught mathematician whose married daughter rejoiced in the name of Haddock. Said mathematician, Olinthus Gregory (a cofounder of London University), did one major experiment in his otherwise unobtrusive life. In 1824 he measured the speed of sound using rockets, cannon, mortars, and bells, in all weather, day and night, over land and water. He concluded intelligently that the direction of the wind affected sound speed, as did temperature, and that echoes were neither slower nor faster. Gregory also translated a work on astronomy by a French scientific cleric (canon of Notre Dame) named R. J. Hauy, otherwise known for his work on crystals and for the fact that his brother Valentine founded what would become the Royal Institute for Blind Children, in Paris. In 1823 a French army captain, Charles Barbier de la Serre, brought the Institute an idea for which he had failed to raise any interest among military personnel, who had taken a typically shortsighted view of the matter.

Barbier's idea made possible writing and reading in the dark, thanks to a script which consisted of patterns of raised dots pricked out in parchment or paper. To read the message in the dark, you ran your fingers over the pattern of dots. Hauy's Institute tried it out on the students, one of whom was an eleven-year-old who came up with several improvements. Three years later, the same student had developed major modifications to the system and made it easier by arranging the dots in groups of six, set in two columns of three. The new version was to become universally known, and named after the boy, as Braille. By 1838 when, thanks to official disinterest, the Institute had fallen on hard times, it was visited by a prominent politician who went back to Parliament and delivered an impassioned appeal for funds. The school was saved because Alphonse de Lamartine was good at passion. Famous for the very public affair in 1816 with the wife of the librarian of the French Academy, four years later Lamartine's first publication of his Romantic poems opened every boudoir door in Paris (he said that love affairs helped him to relax). In 1832 he and his English wife went on a two-year trip to the Middle East, during which he visited the famous eccentric Lady Hester Stanhope, in her fortress on Mount Lebanon, where she had been for twenty years. Lady Stanhope lived in the style of an Arab princess, smoking hash and haranguing her rare visitors with tales of a brilliant youth as

than that of his pal, the duke de La Rochefoucauld, whose death by mob two years earlier showed that not even dukes were exempt from revolutionary justice (it says here). Duke R. was the thinking person's liberal, as well as a constitutional monarchist, brilliant soldier, scion of one of France's great families, and amateur scientist who wrote on botany. His encyclopedic knowledge about decaying vegetation helped when he was briefly made head of the Department of Explosives (compost is an essential ingredient in making gunpowder). As a powerful supporter of the American Revolution, he was also made honorary citizen of New York in 1785, the same year he met Jefferson during the latter's trip to Paris.

With Jefferson came his secretary, William Short, who promptly fell for Rochefoucauld's lovely (and thirty years younger) wife, Rosalie. In short order, Short was having an affair with her. Short's brilliant career then got in the way, as he was shuttled back and forth by the U.S. government from pillar to post (a.k.a. Netherlands, Spain, and Washington) to the extent that Rosalie finally gave up on him. In any case, she wanted to stay in France and didn't fancy becoming plain Mrs. Short, so she married an elderly aristocratic cousin. Short went back to the States, set up house in Philadelphia, made a fortune with investments, and played host to many visiting French *gros légumes* such as Tocqueville, Lafayette, and not least (in 1815) Joseph Bonaparte. Joe had done lots: besides acting as lawyer, author, and Corsican heavy, he'd run France when his brother (you know who) was off fighting the wars. In return for which, he'd been made king of Naples and Spain. When Short met him, with Napoleon in exile and the family royalty business in ruins, Joe was about to become a resident of Bordentown, New Jersey. Here, he played charming host with zee French accent at his palatial white-and-yellow mansion, Point Breeze, surrounded by extensive acres and the bits and pieces a prince picks up in life: Murillo, Velázquez, Titian, Van Dyck, Raphael, Canova, da Vinci, and Rubens, as well as lots of paintings of naked women. A small house in the garden housed his nephew, who had married Joe's daughter.

Charles Lucien Bonaparte was too busy with his hobby to become part of the Bonaparte power play. Shortly after his arrival in the United States in 1823 he joined the Philadelphia Academy of Natural Sciences and started

niece-housekeeper to British Prime Minister Pitt. Three years after Lamartine's visit, the son of a girlhood friend turned up at Stanhope's gate. William Kinglake was on an eighteen-month tour of the Middle East as a break from his socialite life in London and the company of literary luminaries like Thackeray and Tennyson. After a brief sojourn with Lady S., Kinglake returned to civilization, where he was persuaded to write up his journey, and when the book came out in 1844, it was an instant best-seller. A year later the travel bug took Kinglake to French-occupied Algeria, where he spent time with (and wrote about) a local legend, the intrepid Colonel St. Arnaud, leader of the "flying columns," so-called (as was St. Arnaud). In fact he was an ex-actor whose real name was Achille Leroy.

After several years of some (pretty nasty) military action in Algeria, in 1851 St. Arnaud returned to France, and helped with the coup that gave Louis-Napoleon the throne and St. Arnaud a field marshal's baton and command of the French troops in the Crimean War. During this, in September 1854, he won the battle of Alma together with the British commander Lord Raglan. Raglan achieved infamy (some say undeserved) as the man who, one month after Alma, ordered the charge of the Light Brigade, the event where six hundred British cavalrymen misunderstood ambiguous instructions, rode straight into the Russian artillery, and were cut down almost to a horse. What cooked Raglan's goose were the dispatches (published in the London *Times)* written from the front by one of the greatest of all war correspondents, William Russell. Russell roused the paper's influential readership to fury (and caused the government to fall) with his descriptions of the appalling conditions being suffered by British troops in the Crimea. Russell then moved on to even greater bylines, reporting on six more wars: Civil, Austro-Prussian, Franco-Prussian, Zulu, and the Indian Mutiny. During the latter, he got a little too close to the fighting and almost died of his wounds, had it not been for the surgeon's skill and (as Russell reported) use of a new antiseptic, iodine. Originally discovered earlier in the century by a French gunpowder maker, when iodine crystals were serendipitously deposited by his mixture of acid and ash of kelp—a replacement for wood ash, thanks to the general lack of trees, thanks to wooden ships. This use of seaweed in gunpowder gave kelp growers along the poverty-stricken and inhospitable coastlines of Brittany and Scotland a new lease on life. As usual in such cases where a plentiful natural product is discovered, interest grew in what other profitable ways kelp could be exploited.

In 1863 a chemist named Edward Stanford became consultant to the Scottish kelp industry and then manager of the British Seaweed Company. In 1883, after years of study, Stanford discovered that kelp contained algin. Much later, a hydrocolloid derivative of algin (algi-

studying American birds. By 1838 he was back in Europe being a great ornithologist (with a major work comparing the birds of Philadelphia and Rome). Charles Lucien is described as "pompous and difficult," which must be why he became such great buddies with bird painter J. J. Audubon. When they met in 1824 and Charles asked Audubon to do some book pictures, Audubon was so unpleasant to the other illustrators that they refused to work with him. This left Audubon no option but to leave for Europe, where he could play the simple frontiersman and charm the pants off everybody and get his paintings of American birds published. Great if you like chocolate-box art. Audubon was more of an artist and Frenchman (born in French Santo Domingo) than he was a researcher and English speaker, so on a visit to Scotland in 1830, he hired a young local naturalist, William MacGillivray, to check his grammar and punctuation and to provide zoological detail for the picture captions. The two of them worked together, on and off, for nine years on Audubon's mammoth *Ornithological Biography.*

MacGillivray had spent his teens walking all over the Scottish Highlands and islands, shooting things so he could dissect them, and for this reason was well prepared for his two great life efforts. First: walking 837 miles (with notebook in hand) from Aberdeen to London, in order to visit the British Museum bird collection. Second: his (the first) classification of birds by anatomy rather than plumage. Somewhere in all this walking back and forth, MacGillivray found time to have thirteen children. The eldest of which, John, followed in his father's footsteps (not literally) to become a naturalist on three long exploration voyages: the first in 1842 to the Australian coast and the Great Barrier Reef and a third in 1852 to the South Pacific, during which he was dropped off in Sydney (for "intemperate habits") and stayed in Australia for the rest of his life. The second voyage (in 1846, to Australia) was on HMS *Rattlesnake.*

One of John's shipboard companions was Thomas Henry Huxley, the vessel's assistant surgeon and hypochondriac. Huxley spent the entire trip dredging up stuff and discovering a new group of marine organisms for which he got much kudos (membership in the Royal Society). He also became a jellyfish guru, and wrote on squid, animal individuality, skulls, and fossil horses. In 1859 he got an advance copy of Darwin's opus and became the Great Defender of

TRACK ONE

nate) became a major ingredient in the manufacture of everything from printing inks and ice cream to dental-impression molds. In this last case, the most attractive feature of alginates was that the material didn't stick, so the cast was easily removed once the mold had set. It was this particular property that, in the 1930s, attracted the attention of a Hungarian ophthalmologist named Joseph Dallos.

END TRACK ONE

TRACK TWO

Evolution. Known as "Darwin's Bulldog," Huxley spoke to sold-out lectures all over the United States. And then at less crowded venues, such as the Metaphysical Club in London, he spoke on subjects including: "Has a frog a soul?" At a Royal Institution lecture in 1856, Huxley heard Institution Professor John Tyndall (like Huxley, a popularizer of science) raise an exciting possibility: that the way pressure on crystalline materials caused slatey cleavage might relate to glacier structure. In true Victorian fashion, Tyndall and Huxley decided to head for the Alps to take a closer look. Tyndall then discovered glacier flow and a passion for mountaineering and was one of the earliest up the Matterhorn.

His Alpine first consisted of spending a night on the summit of Mt. Blanc in 1859 with a longtime pal, chemist Edward Frankland, who used the occasion to reveal that candles burned at the same rate up high as down low but gave out less light. This finding led Frankland to study the effects of atmospheric pressure on combustion (a matter of interest to gaslight manufacturers and to military people lighting fuses at altitude). In 1865 Frankland also helped to set up another mountain-top experiment with his Zurich physiologist brother-in-law, Adolf Fick, who climbed a Swiss Alp (the Faulhorn, exactly 6,417 feet high). Having previously followed a protein-free diet, Fick regularly measured the amount of nitrogen in his urine as he climbed, proving by this method that protein, not fat, was the source of muscular energy. Adolf's ophthalmologist nephew of the same name set his sights lower, concentrating on the kind of astigmatism known as keratoconus (in which the cornea grows outward like a cone), which couldn't be corrected by eyeglasses. Young Adolf's idea was to replace the cornea with a clear shell of glass. In 1888 he tried one on himself. This revealed a problem: how to make sure it fit closely enough to work.

END TRACK TWO

AND FINALLY . . .

Joseph Dallos used alginate to make an easily detachable eye mold which could then be used to shape Adolf Fick's contact lenses, so that they would fit well enough to be worn without discomfort.

SEVENTEEN

1742: BOW STREET, LONDON, TO BAR CODE

Eighteenth-century Bow Street, London, was a rough neighborhood, full of taverns and brothels and criminals who'd slit your throat for a shilling. As a result, it was also where the first police force was formed. Bow Street is also just round the corner from Drury Lane Theater, which was why, in 1742, three of the most famous actors in England lived at No. 6.

TRACK ONE

Charles Macklin lived in the upstairs flat with his wife and child. A year earlier, he'd had his first success with a performance of Shylock in Shakespeare's *The Merchant of Venice,* and over the next fifty-six years he would rarely be off the stage. Arrogant, narrow-minded, and violent, Macklin had killed a fellow actor in 1735 in a row over a wig, and conducted his own (successful) defense. His life was filled with duels (never fought) and lawsuits. He played every role from slapstick to high tragedy, and at the age of eighty-five gave his farewell performance once again in the role of Shylock, after which he lived to an eccentric ninety, bathing only in gin. In the apartment downstairs lived the period's most famous actress, Peg Woffington, described by Dr. Johnson as "dangerously seductive." In 1742 she was the partner (both ways) of the most famous actor alive, David Garrick.

Woffington had started life selling vegetables on the streets of Dublin, and then in 1740 she took London by storm in the role of Sir Harry Wildair, when she showed enough leg to cause a riot. Painted by Reynolds, Woffington was adored by everyone, played every major female Shakespearean part, and broke with Garrick in 1744 after a row on the day they were due to be married. Woffington moved in with an aristocrat lover, then took her own house at Teddington, a fashionable village outside London, where she became a parishioner of the Reverend Stephen Hales, vicar of St. Mary's, who was the most famous English scientist of the period. Much of what Hales got up to can't be described for readers with a delicate constitution. Apart from sticking glass tubes into plants to measure the power of rising sap, he did similar things to animals. I'll say no more. Like everybody else, Hales was captivated by respiration (what was air?) and physiology (how did muscles work?). It was the muscle question that got him into sap, looking to see if a human equivalent of sap was what made muscles contract and expand. His investigation of air involved vacuum pumps and testing how long he could breathe his own exhaled air without serious consequences (about a minute). Bad air was clearly bad for you, so Hales came up with a ventilator for slave ships and prisons (plenty of bad air in both).

In 1751 a favorable report on the use of his ventilator aboard a slave ship came from owner Henry Ellis, who'd previously sailed on yet another of those searches for the Northwest Passage and spent so much time transatlantic he'd become an American expert. In 1757 he arrived in Georgia as governor, and he busied himself cultivating silk; shipping home seeds, flora, and fauna; corresponding with Linnaeus; and naming a shrub after Hales. He also did the governor's job so well he was appointed to the (absentee) post of governor of

TRACK TWO

In the downstairs apartment of No. 6 lived David Garrick, the man who introduced modern techniques to stage acting. In 1741 Garrick's first major role was Richard III, and he was an overnight sensation. For thirty-five years, this small, fidgety man with the expressive face ruled the English stage in parts ranging from low comedy to King Lear. His extraordinary talents brought him into contact with everybody who mattered: artists, writers, politicians, scholars, and royalty. In 1747 he took over the management of Drury Lane Theater, where he introduced unheard-of reforms such as shutting the theater doors after a play started and ending discounts for people who came late. In 1771 he met the man who would do for stage design what Garrick was doing for acting. Philippe Jacques de Loutherbourg, an established artist, joined Garrick and radically transformed the theatrical experience with on-stage innovations like burning houses, waterfalls, fog, atmospheric lighting changes, lush backdrops, falling walls, heaving oceans, sound effects, and puppets. He accomplished all this while an active member of the Royal Academy and doing serious landscapes and book illustrations.

In 1786 Loutherbourg and his wife left London for Switzerland, where he lost all his money, returned destitute, and temporarily became a faith healer. This switch to the mystical was perhaps triggered by the influence of his recent traveling companion, a man who was an expert in a different kind of illusion: "Count" Cagliostro, international con man. Born Giuseppe Balsamo, after an apprenticeship among petty criminals, the Sicilian Cagliostro went into the magic-elixirs-and-séances flimflam and began to take serious amounts of money from gullible aristocrats all over Europe. The hook for his credulous and greedy clientele was Cagliostro's claim to be able to turn base metal into gold. By 1773 Cagliostro was top of the bill at Versailles, until he got out of his depth in a scam involving a courtier, Queen Marie Antoinette, and a diamond necklace. As a result of which, Cagliostro spent six months in the Bastille and then left for London (and eventually Switzerland, together with the Loutherbourgs).

In the end, Cagliostro's fatal error was Freemasonry, which was useful in northern Europe but dangerous in Rome—where the Inquisition didn't go for funny hand-

Nova Scotia, then retired early on a handsome inheritance. From 1757 till 1761 his Georgia provost marshal was William Knox, who then became British agent in Georgia and Florida (during which his recommendation that the colonists have seats in Parliament was rejected). When the War of Independence started, he was Undersecretary of State for American affairs and, after the war, failed to set up a colony in Maine for Loyalist Americans who'd fought for the Brits. In the end, he managed to create New Brunswick, and many Loyalists went there. Earlier, in 1769, Knox had written a piece reacting to a pamphlet on America written by an American doctor living in London. The physician, Edward Bancroft, was to become the spy of the century. After spending time in Suriname, he settled in London and his cosmopolitan ways won over bon vivants like Ben Franklin. In 1776 he was recruited to spy on the Brits for Silas Deane, at that moment in Paris arranging a Franco-American treaty. At the same time, Bancroft was also recruited by an old Suriname pal, Paul Wentworth (now head of the British Secret Service), to spy on the Americans in Paris.

Wentworth, ultimate schmoozer and fluent in French, ran the British European espionage network, and when it became clear that the French were going to arrange an anti-British deal with the Yanks, he had the brilliant idea of offering the U.S. Congress thirty million pounds and two hundred aristocratic titles if they'd all become Loyalists. No deal. Wentworth and Bancroft (and, unknown to Wentworth, Deane) were each using their advance know-how about the imminent conflict to manipulate the stock market to their advantage. And when it was all over, and Deane was on the way home to squeal to Congress about Bancroft, Deane died in mysterious circumstances.

Meanwhile it was all invisible ink and leaked memoranda, dead-dropped in empty bottles (in tree trunks in Paris), which were then collected and delivered to the British ambassador, Lord Stormont. Stormont was also keeping a close eye on two English iron-making brothers, John and William Wilkinson. In 1774 John had perfected a system for boring out cannon barrels from a solid piece of cast iron. This meant the guns were less liable to explode when being fired and so of great interest to the French, who were about to export munitions to help the Americans. In 1775 a French general visited the Wilkinson plant, and in 1776 William arrived in France, where, with materials smuggled from England, he set up a foundry and cannon-boring machine (the company, Le Creusot, still exists today, and makes kitchen hardware).

In Britain, Wilkinson's boring machine was so good it kicked off the Industrial Revolution, because it was just what James Watt and his partner Matthew Boulton needed to make the steam engine work

shakes and put Cagliostro in prison for life. At one point, while passing through southern France, Cagliostro met the only Freemason who could equal him in braggadocio and venality: Giovanni Giacomo Casanova, about whom it must be said that he probably did everything he claimed. This was especially true when it came to seducing (by his own calculation) 122 women (including his own illegitimate daughter) and then leaving via the back stairs, which he did once every few months for most of his life. Soldier, adventurer, diplomat, spy (France, Venetian Republic), writer, and librarian, Casanova made a million (on the Paris lottery, which he set up) and spent it all on fun and doctors' bills (venereal disease). In 1797, in Prague, he helped Mozart revise the libretto of *Don Giovanni,* written by another pal to whom Casanova had recounted some of his adventures and probably acted as the title-role model. The pal, Lorenzo da Ponte, also did Mozart's librettos for *Così fan tutte* and *The Marriage of Figaro.* He'd started as a priest, and then in 1781 worked for Mozart's archrival Salieri at the Vienna Opera House till intrigue forced him to London and heavy debts. To avoid these, in 1804, he emigrated to the United States and settled in New York, where for thirty-three years he scraped along as a grocer, bookseller, and professor of Italian at Columbia. It seems to have been a bookstore encounter that got him the teaching job.

The customer was C. C. Moore, recent Columbia alumnus and son of the Columbia president. Moore was professor of Hebrew and Greek literature at the General Theological Seminary in Manhattan, and wrote a major Hebrew-English dictionary as well as a biography of the Albanian patriot Skanderbeg. In spite of which, he is remembered for also having written (in 1821), " 'Twas the night before Christmas." Originally, two of the reindeer were called Donder and Blixem, but their names were altered to the more familiar names Donner and Blitzen by the editor of the 1837 *New York Book of Poetry,* Charles Fenno Hoffman, lawyer turned would-be author. Although he did well with his account of a year spent traveling the Wild West (at that time Pennsylvania, Ohio, Michigan, Indiana, Missouri, Kentucky, and Virginia, Hoffman's novels never took off. He was also clobbered by British reviewers for being so derivative as to verge on plagiarism. Hoffman did pretty well, however, as an editor of other people's efforts in various literary mags, and in 1840 he became associate editor of Horace Greeley's

efficiently. Wilkinson bored out a thin-walled cylinder, nine feet long and six feet in diameter, so accurately that Watt was able to get a good seal (and therefore a good head of steam). In 1803 Watt and Boulton were approached by a former artist turned engineer, Robert Fulton, for an engine to power his new steamboat. The technology-transfer law was circumvented once Fulton agreed to stop building submarines for the French, so the engine duly arrived in America. There, by 1808, Fulton and his partner Livingstone had a commercial service up and running on the Hudson, and a thirty-year New York State monopoly to continue doing so. More than twenty would-be competitors took the pair to court for violating a federal law that regulated interstate commerce. New York Supreme Court Chief Justice James Kent held that in this case the state was sovereign, and since Congress hadn't passed a law specifically banning that kind of activity, the New York monopoly stood. After Kent retired in 1823, he did lucrative consultancy work for New York City and the Cherokee Nation.

He also joined a new lunch club, the *Bread and Cheese,* recently founded by a young writer from upstate New York who had just had his first major success with an American Revolution novel: *The Spy.* In 1826 James Fenimore Cooper then made enough money out of *The Last of the Mohicans* to take his family to Europe for five years and to write himself into the role of respected American national novelist (making friends with Lafayette while he was in France didn't hurt). One of the other members of Cooper's little New York club was Asher Durand, who since 1818 had been studying drawing at the Academy of Fine Arts and became good enough to do banknotes and then to open a printmaking shop. After he gave up engraving and turned to canvas, Durand effectively invented the Hudson River School style with *View Towards the Hudson Valley* (1851) and *In the Woods* (1855). He then wrote *Letters on Landscape Painting* that told wanna-bes how to do it. His greatest engraving work was probably *Ariadne,* inspired by the "indecent" nude painting by John Vanderlyn, who spent half his career painting or copying portraits (of Washington, Fulton, Aaron Burr, and other movers and shakers) and made enough money so he could paint historical subjects that didn't sell.

Like everybody, Vanderlyn spent time in Europe, until a couple of engravings of Niagara Falls, which he'd done years before and left behind with an agent, suddenly made enough for him to return home. More portraits followed, while he tried (and failed) to run a cyclorama (the audience sat and looked at gigantic paintings of places like Versailles) in an art gallery named the New York Rotunda. It was in the other Rotunda, on Capitol Hill, that Vanderlyn finally made the big time, with a commission to fill one of the remaining

New Yorker. A year later the *New Yorker* became the *Tribune* and one of America's great newspapers.

Greeley's editorials led the fight against slavery, and the mistreatment of Native Americans and promoted the Western movement ("Go West, young man"). Greeley helped found the Republican Party, supported Lincoln's nomination, lobbied for a transcontinental railroad, and wrote *Recollections of a Busy Life,* one of the best of American autobiographies. His only major failure was to lose the 1872 presidential election. But he succeeded in awaking the nation's conscience to many of the issues of the day—among them the position of women, to which end he hired Margaret Fuller in 1844, making her the first woman in the working press. The following year, her *Women in the Nineteenth Century* called for equality between the sexes and struck a chord. The book was sold out in two weeks. In 1846 Greeley sent Fuller to file stories from Europe. In Paris she met one of her idols, the French female novelist and feminist George Sand. The following year, Fuller was again smitten, this time in Rome, by the Marquis of Ossoli and (probably) married him. After living through the French invasion of Rome in 1848 (during the struggle for Italian unification the marquis fought and Fuller nursed the soldiers), they both escaped to Florence, and Fuller wrote up her experiences for the *Tribune.* In 1850 they and their baby left Livorno for New York, and in a gale the ship foundered within sight of the American shore, and they all drowned.

Their last evening in Florence had been with their new friends Robert and Elizabeth Browning, English poets (she better than he). Elizabeth had been writing for some years before meeting (and eloping to Italy with) Robert in 1846. Robert was still unknown when they returned to London for a short stay in 1851, which was when they met admirers of Elizabeth's work: stars like Tennyson, Kingsley, and Carlyle—and art critic John Ruskin, whose work Elizabeth rated highly. Ruskin more or less established the school of art criticism, beginning with his 1843 *Modern Painters.* He went on to concentrate on architecture with later classics like the 1853 *Stones of Venice.* He was hooked on Gothic, considering medieval architecture the highest form of expression because it had been done by ordinary craftsmen who were happy in their work (unlike contemporary factory workers whose cause the socialist Ruskin espoused). So he spoke and wrote reams on how the Gothic Revival (and the

TRACK ONE

panels (with *The Landing of Columbus).* One of the painters he beat to the job was Samuel Morse, who was so ticked off about losing that, in 1836, he gave up art for an idea he'd been working on for several years: electric telegraphy. And on May 24, 1844, Morse achieved immortality with his famous message (transmitted from Baltimore to Washington), tapped out, before an astonished Congress, in Morse code.

END TRACK ONE

TRACK TWO

hundreds of pseudo-Gothic buildings going up all over the country) was good for the moral tone of Great Britain. This slightly mystical view of things also expressed itself in his membership in the Society for Psychic Research, which was deeply into table rapping, trances, mediums, and such.

All were considered respectable activities, as attested by the Society's scientific adherents, which included Crookes (cathode rays), J. J. Thomson (electrons), Wallace (evolution), and Oliver Lodge (in 1881 professor of physics at University College, Liverpool), coinventor of the coherer, a small glass tube filled with iron filings. When a very weak electrical signal arrived, the filings would cohere, marking the arrival of weak signals such as radio. Early in the twentieth century, amplification of weak signals became vital. So Yale graduate Lee De Forest played around with the "Edison effect": in a vacuum tube, heat a cathode that is coated with a material that gives off electrons when heated. The negative electrons rush in a powerful stream to the positive metal baseplate. De Forest stuck a tiny grid between cathode and baseplate and charged it with the incoming weak signals to be amplified. These modified the powerful flow of electrons, and the baseplate gave off a matching but much amplified signal. De Forest then found that if the signals came from a loudspeaker, they could be used to vary a lightbulb's brightness. This varying light would varyingly expose a moving film negative. You then developed the film, ran it through a projector, and shone a steady light at it. Varying amounts of the light would get through the film and hit a light-sensitive cell that generated a matching, varying pattern of signals. These in turn generated sound from a loudspeaker, loud enough for the sounds to be heard in a cinema—thus creating the talkies.

END TRACK TWO

AND FINALLY . . .

In 1952 Americans Norman Woodland and Bernard Silver drew Morse-code words as a series of thick (dash) and thin (dot) vertical lines on a white background, shone a light at the pattern, and echoing the work of De Forest, captured the bounce-back on a light-sensitive cell—thus creating the bar code.

EIGHTEEN

1739: THE GRAND TOUR TO LIQUID CRYSTAL DISPLAY

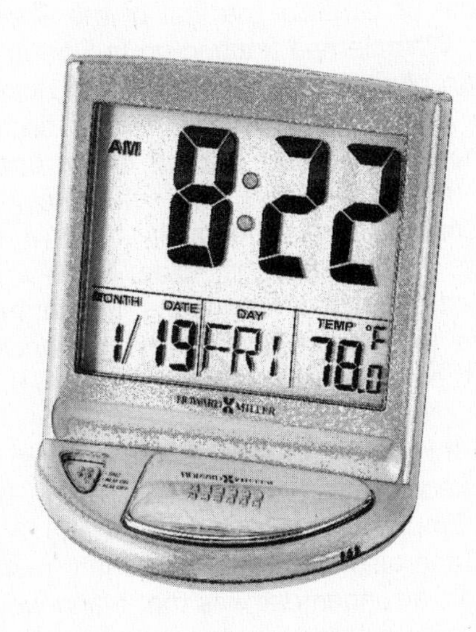

In the mid eighteenth century, the fashionable thing for young British aristocrats was to head off to the Continent with a tutor and get some Culture. After which they were then supposed to come home, settle down, and forget it all. This Grand Tour, as it was called, could last several years, depending on how much fun was being had. Some people never came back.

TRACK ONE

Horace Walpole set off on his Grand Tour in 1739. Walpole, rich, the latest scion of a long line of elite, and son of a prime minister, easily met all the Grand Tour has-money-needs-culture requirements. Unlike others, however, Walpole was no aristocratic lout but already a scribbler and chattering-class novitiate. Ten years after he got back, he would turn out to be the man who almost single-handedly kicked off the English Romantic movement in 1747 by turning a modest residence (at Strawberry Hill) outside London into "Strawberry Hill Gothic": imitation medieval everything, filled with knickknacks, and a printing press from which he issued essays and chirpy notes on what was hot and what was not. In 1764 his own nouvelle-vague contribution was *The Castle of Otranto,* the first of the Gothic horror novels. While on his tour, Walpole had fetched up in Florence at the welcoming residence of an old pal, the British minister Horace Mann, whose forty-six years in Italy were filled with visiting firemen, art dealing, and spying for the British Secret Service. Mann operated a network that kept round-the-clock watch on the ex-Royal family Stuarts, James and son Charles, who sequentially tried (and failed) to grab the English throne. Cut off from Jolly Olde, Mann wrote thousands of letters home, filled with now-fascinating Florentine social chitchat trivia, while amassing a pile of newly excavated ancient Roman carvings and offering them for sale at inflated prices to Brits now crazy for anything classical.

His partner in this profitable little scam (they did all their deals by mail) was a well-placed Italian cardinal (nephew of the pope and the Austro-Hungarian Empire's representative in Italy), Alessandro Albani, who in his spare time also acted as Mann's informant on those plotting Stuarts. The art arrangement was that Mann would find the buyers, while Albani would provide the goods and get them smuggled out of the country. Albani had been "collecting" and selling antiquities for years (to Polish royalty, among others), and by 1755 had also kept enough aside for himself to require a large custom-built villa (neoclassical, of course) to house the stuff.

One of the guys who worked on the villa décor was a visiting young French architect, Charles-Louis Clerisseau, whose later claim to fame would be his work in 1785 with the former Ambassador to France, Thomas Jefferson, designing the Virginia State Capitol on the model of the Nimes Roman temple known as the *Maison Carrée.* When Clerisseau met Albani he'd just got back from a trip to Splalato (now Split, Croatia) to sketch the Emperor Diocletian's palace with his traveling companion and ancient-architecture buff, Robert Adam, who would later reproduce the drawings (without attribution, of course) in

TRACK TWO

Horace Walpole's companion (in more senses than one) on the Grand Tour of 1739 was Thomas Gray. They'd been students together at Eton and then Cambridge. Gray was the consummate esthete, writing delicately balanced Latin poems and equally delicate English verses. His most famous effort, "Elegy in a Country Courtyard," is probably the best-known eighteenth-century poem and displays everything he stood for: decorum, polish, an elegant mixture of classical-past order and Romantic-to-come sentiment. It is a quiet piece. And "quiet" is the word to describe most of his college-bound life: that of a rich gentleman with nothing more demanding to do than read and scribble. This scholastic existence was disturbed late in life by the brief sleepover in 1769 of a young, handsome, bisexual Swiss named Bonstetten, about whom Gray went discreetly gaga. Bonstetten had tested the waters in Florence, dallying with Louise von Stolberg, wife of the drunken Bonnie Prince Charlie, but eventually came down on the side of gaiety—as had Gray. After Cambridge (and numerous passionate letters from Gray), Bonstetten spent a couple of years administering Italian Switzerland, and then, after Napoleon's 1798 invasion, he moved permanently to Geneva and the crowd round the lake (Voltaire, de Staël, Necker, Constant, et al.). Here, he wrote a major work on the climatic reasons for the differences between northerners and southerners.

One of Bonstetten's other amorous correspondents was Johan von Muller, whom he'd met just after getting back from Cambridge. Muller was the big history fish in the small Swiss pond and was (unusual for the times) openly gay. In 1771 he began his great *History of Switzerland*. Muller's life was a series of librarianships, culminating in the Imperial Library, in Vienna, where, in 1802, he fell in love by mail with a young Hungarian duke invented by a kid Muller was teaching at the time. Meetings across Europe were arranged with the fictitious duke, who was always "expecting an inheritance," and only needed a temporary loan each time, but never turned up for any meetings. Muller eventually ran out of money. In spite of which, his *History of Switzerland* reached year 1489. This was just as well for Friedrich Schiller, who took the lead character of his play *William Tell* from Muller's history. Schiller was another historian but, fortunately for posterity and the theater, met Goethe and

a big book. Adam was out from Britain to learn all he could about what might eventually make him the king of chic back home. His venture paid off. After two years in Rome, measuring the ruins and peering over Piranesi's shoulder, in 1759 Adam arrived back in the United Kingdom. Five years (and that big book) later, any Brit gent with more money than sense and with conspicuous expenditure in mind couldn't get enough of Adam's porticos, Etruscan dressing rooms, tunnel-vault libraries, and peristyle entrance halls, all complete with matching-in-every-detail Roman décor—even the commodes.

Adam's sister Mary was married to John Drysdale (a suitably named bore and, in 1773, principal clerk to the Assembly of the Church of Scotland), about whom it can only be said that he went to school with Adam Smith and his sermons were reported to be fire and brimstone. His son-in-law, Andrew Dalziel, edited two volumes of the sermons (with a fawning preface) and took over from him in the General Assembly. Dalziel made the most of what he had. Son of a carpenter, by the late 1770s he was also Edinburgh Greek professor, pal of such heavies as Hutton (geology), Smith (economics), Black (chemistry), and Cullen (medicine), and in 1783 helped found the Edinburgh Royal Society. In terms of his lasting effect on history, though, you might say he rose without trace. In 1786 his class included a precocious (first novel at age twelve) seventeen-year-old Swiss, Benjamin Constant, visiting Edinburgh to enjoy the stimulus of the Scottish Enlightenment as part of a reverse version of the Grand Tour. By the age of twenty-nine, Constant was in Paris being a well-known political commentator—an extremely dangerous profession, given the revolutionary fervor of the moment. Constant's companion was no less risky: Germaine de Staël, queen of the salons, was another first-work-at-twelve author and soon-to-be mother of Constant's child. After her tongue got her exiled from France, Constant went, too. When things intimate inevitably went wrong, he married a German noblewoman but hedged his bets for two more years by continuing with de Staël. After the usual ins-and-outs with Napoleon (emperor, deposed, emperor, deposed), Constant took up the cause of representative democracy, separation of powers, and a bill of rights, and died a French national hero.

De Staël did equally well, at one point sleeping with both Constant and German writer August von Schlegel, writing about everything (according to Goethe, whether she understood it or not), and generally becoming the bedroom everybody wanted to visit, in both senses of the term (she also conducted her salons there). She picked up Schlegel on the same visit to Weimar during which she wrote a quickie on German Romanticism (she coined the term), and also met and transfixed Friedrich von Schelling. He was the guy who codified

was persuaded to take drama more seriously. From 1799 he lived in Weimar, Germany (with Goethe and the cream of German literary talent) and turned out great stuff including *Mary Stuart* and *Tell.* In tune with the revolutionary times, his material was all about individual freedom and political-moral responsibility. It was good enough to be used by Verdi (for *Luisa Miller),* Beethoven ("Ode to Joy," a.k.a. the choral bit in the Ninth Symphony), and above all, by Rossini (1829, *William Tell). Tell* was the last of Rossini's operas before he retired, is extremely long, and is very seldom done in its entirety (some feel with good reason). It's said that when told the Paris Opera was performing only the second act, Rossini said "All of it?" *Tell* came at the end of a prolific career and thirty-four operas, starting back in 1813 with *Tancredi,* and including some of the most popular and widely known works ever written (especially the *Tell* overture). At the height of his fame, Rossini was bigger than life, loved food ("Give me a menu and I'll set it to music"), and inspired the dish *tournedos Rossini* (try only when ravenous).

In 1825, not long before *Tell,* Rossini gave singing lessons to a teenage Brit, Henry Russell, who'd been on stage since the age of three. After some chorus work in London, in 1835 Russell took off for North America and became a smash hit with what were effectively the first-ever pop songs. Russell did nonstop one-night stands all over the country with his one-man show. His material was what we would now call "tear jerk": mothers dying in old armchairs, emigrants' farewells, deserted wives and children, and over 250 others, including "Woodman, Spare That Tree," "Ole Dan Tucker," and most famous of all (in 1838) "A Life on the Ocean Wave," which became the British Royal Marines' regimental march. This last owed its words to Epes Sargent, who, with Russell, spent time that year with Fenimore Cooper and dedicated the song to him. Sargent passed most of his life as an editor for magazines such as the *Boston Daily Advertiser* and the *New York Mirror.* He was also into table rapping, clairvoyance, and school textbooks (where he made his money). In 1839 he wrote one of his two dramas, *Velasco,* considered worth a mention seven years later in Edgar Allan Poe's "Literati" column. It was one of the rare Poe reviews that wasn't a hatchet job.

Life for Poe was one long hatchet job: he was in turn depressive, drunk (he would die of alcoholic poisoning),

the Romantic belief that there was some fundamental substrate that unified all nature and humankind (a Grand Unified Theory before the Grand Unified Theory). Schelling called it *Naturphilosophie* (with a name like that, it couldn't fail). All progress was due to opposite poles of Schelling's substrate clashing and resolving. This got physicists looking at the concept to useful effect when it led to electromagnetism.

Schelling's great pal took these ideas a great deal further. Hegel said the "resolution" thing applied to all history. The great dialectic (thesis to antithesis to synthesis, which then became the next thesis, kicking off the next clash and resolution) was the process behind everything that ever happened. This may be a slight misinterpretation of Hegel, but such is not unusual. Marx took him to mean the class struggle. A young German academic latched onto a different bit of Hegelian thought: that understanding contemporary conditions was essential to historical analysis. Wilhelm Roscher took this relativist approach to economics. There was no "law" of economics, he said. Only different patterns, given different conditions, in different places, and different times. Economic behavior was strictly seat-of-the-pants, relevant to the locale. So state intervention was OK because state bureaucrats knew the problems that needed solving and could do effective things (like welfare).

This idea warmed up a visiting American, Richard Ely, who studied under a Roscher sidekick and, in 1881, took the message (activist, socialist, interventionist) back to Johns Hopkins University. There, with the United States going through an uncertain acceptance of union activity, it got him into big trouble with the industrialist university funders and into one of those vicious academic struggles-for-the-jugular with a fellow faculty member. Simon Newcomb was mathematics and astronomy professor, with a senior position with the U.S. Navy's Almanac Office in D.C. and a string of books and articles on how to apply scientific (and in particular mathematical) principles to economics. By 1892 Newcomb had won, and Ely left for a warmer welcome at the University of Wisconsin. Back in 1879, Newcomb's assistant in the Almanac Office was a young physicist, Albert Michelson, who in 1887 proved the nonexistence of "ether," the mystery, ubiquitous substance through which electromagnetic waves were supposed to travel. Michelson split a light beam and sent one half "upstream" into the ether, through which the Earth was supposed to be moving (so the beam should be slowed down), and the other half at right angles (so there would be no slow-down). When recombined, the two beams were still in phase—so, no ether.

Heinrich Hertz had started all the fuss back in 1886 by discovering electromagnetic waves in the first place. Two years later, Hertz

broke, or widowed. Small wonder that his work fringed on the insane. Still, it blew the socks off such talents as Debussy and Rachmaninoff, both of whom used Poe's material. A few years later, Poe's work was being enthusiastically translated and reviewed by a young French poet not entirely unlike Poe. Charles Baudelaire was a suicidal, syphilitic spendthrift, who used drugs when he ran short of poetic inspiration. A major work entitled *The Flowers of Evil* won him a prosecution for offending public morality, as well as a fine, censorship of his work, and lousy critiques from reviewers unable to take Baudelaire's satanic and lesbian themes. Much of his stress might have come from not being able to decide which of three mistresses he wanted to live with or to marry. The most interesting one was probably Apollonie Sabatier. By 1842 she was a professional model and sleeping partner of several painters. In 1846 a rich Belgian industrialist set her up as his mistress in a Paris house, and her social whirl began. Hostess of the most brilliant salon in Paris (funded by the Belgian), Sabatier was surrounded by glitterati such as Balzac, Flaubert, Lamartine, and Manet. A year later, the Belgian commissioned a statue of her *(Woman Stung by a Snake)* that shocked even the Parisians, because it was clear that what was being experienced had nothing to do with snakebite. In 1860 Sabatier's Belgian went back to his wife, and Sabatier briefly moved on, undeterred, to greater things, a.k.a. Sir Richard Wallace.

Illegitimate son of the Marquis of Hertford, Wallace was loaded, already a connoisseur of paintings, and had a twenty-year-old son by his permanent mistress, Julie Castelnau (which seemed to make no difference to Sabatier). Wallace lived in Paris, and in 1871 he inherited his father's fabulous art, to which he added small purchases such as the greatest armor collection in the world and various major Renaissance works. By the time his widow died and left it to Britain, it had become what we now know as the Wallace Collection. In 1878 Wallace was appointed a British commissioner to the Paris Exhibition, at which the star attraction was the seventeen-foot-high head of the Statue of Liberty, being built at the time by Frédéric Bartholdi with the assistance of Gustav Eiffel. The name of the game was to make the biggest statue ever built (Bartholdi was a freak for the colossal) and to place it in New York harbor. The link with a stable democracy (the

TRACK ONE

left Karlsruhe, to be replaced by Otto Lehmann, an expert in the fine structure of matter as seen through a microscope—just the guy to solve the problem of Friedrich Reinitzer, botany lecturer in Prague, who wrote to him about something that seemed incredible. While investigating cholesterol in plants (such is research), Reinitzer had discovered that between 144.5° C and 178.8° C, funny things happened to cholesteryl benzoate. First, it melted to an opaque fluid, then, at the upper temperature, the liquid went clear. While cooling, it exhibited iridescent colors and then became a white solid. Lehmann peered at the stuff and announced that Reinitzer had discovered a fourth state of matter: not solid, liquid, or gas, but liquid and crystal at the same time. The cholesterol was behaving as if it were crystalline (hence the funny colors, when it did things to the ambient light) and also liquid, as it flowed.

END TRACK ONE

TRACK TWO

United States) would be good for not-so-stable France and concomitantly would serve as a permanent reminder to the United States of French help during the War of Independence. *Liberty Enlightening the World,* facing back toward France as she did, would drive this last point home. The French reckoned without American savvy. When the poem was chosen for the pedestal, it totally changed the statue's nature, making "Miss Liberty" the American hostess welcoming all those "huddled masses, yearning to breathe free." The poet was herself deeply involved with new immigrants, especially those fleeing the pogroms in Russia and Germany.

By 1884, when she wrote the "Liberty" poem, Emma Lazarus was an established writer and (a decade before Zionism) the driving force for the establishment of a Jewish state in Palestine, having traveled to England and met Arthur Balfour. Balfour would later be Prime Minister and then, in 1916, the Foreign Secretary who wrote a letter to Baron Rothschild expressing support (the letter became known as the Balfour Declaration) for the eventual establishment of a sovereign Israel. Balfour's sister Evelyn was married to a landed aristocrat named Strutt, who in time inherited the title of Lord Rayleigh and a fancy English manor house to go with it. Rayleigh was the last of the great science polymaths, covering as he did the whole field of "classical" physics, as well as discovering argon, and getting a Nobel in 1904. Rayleigh ended up as president of every important science society you could name; recipient of double-digit honorary doctorates, professorships, and board memberships of advisory councils; and wrote 430 research papers. In the lab that he added to his manor house, he also solved the mystery of why the sky is blue (reason: light scattering).

END TRACK TWO

AND FINALLY . . .

In 1968 RCA produced the first digital clock, using a technique that brought together Reinitzer's liquid crystals and Lord Rayleigh's light-scattering effect. When an electric current was applied to liquid crystals sandwiched between two thin layers of glass, they turned cloudy and scattered light. The area within the liquid crystal display where this was happening would then become visible as 10:15 (or whatever time it was).

NINETEEN

1795: MAN IN THE IRON MASK TO HOVERCRAFT

Did Louis XIV secretly put his twin brother in prison and make him wear a metal face cover until he died in 1703? *The Man in the Iron Mask* mystery fascinated the eighteenth century (and, later, Hollywood). In 1795 it also inspired Anne Yearsley to write *The Royal Captives.* No big deal. An average work. However, Yearsley (pen name "Lactilla"), mother of six in Bristol, England, was self-taught and a milkmaid. Not the kind of person likely to be able to write, let alone write novels. Or poetry. Or publish any of them. That she did so was in great part thanks to two of the most talkative women you'd never hope to meet.

In 1784, Yearsley was a cook's helper to Hannah More, who took her up as a shining example of what the lower orders could do if given the opportunity. At the time, high-profile queen-of-ladies-who-lunch, More was writing her own glitzy review *(Bas Bleu)* of who was in and who was out in the London salons, where those with pretensions to culture made and unmade reputations over tea. More saw herself as a libertarian. She promoted a modicum of education for the poor (made them better servants), help for destitute children (inculcated obedience to authority), and classes for women (taught them the art of conversation). She was also against slavery (only outside England), after she'd met a young charmer, the tiny (five feet four inches) William Wilberforce. All around him Wilberforce saw corruption, boozing, adultery, obscenity, and all the other things well-heeled people did (Wilberforce too, until he saw the light). So he took up the cause of blacks, and over twenty years he managed to get slavery outlawed. On the way, in 1791, he helped found the Sierra Leone Society to establish a settlement in West Africa in which "returning" slaves could live free. This dubious go-home sentiment was shared by an unusual black American entrepreneur whom Wilberforce met in London. Paul Cuffe was a Massachusetts shipowner who visited Sierra Leone twice, carrying African American settlers. In 1816 he was in the United States, helping to start the American Colonization Society. In 1821 the Society bought a tract of land on the West African coast that would later become the sovereign state of Liberia, which would, during the nineteenth century, receive up to twenty thousand repatriated former slaves.

This move-them-out approach to antislavery (as opposed to "remain in America, black and free") attracted support from white Americans. In 1821 Quaker Benjamin Lundy started a newspaper with the unambiguous name *The Genius of Universal Emancipation,* started networking all the antislavery groups, and pushed for more black "colonies" to be set up in Canada, Haiti, and (Mexican) Texas. In 1827 the newspaper got a scoop that was too good to ignore. A few miles upriver from Memphis, Tennessee, at one of the new, interracial utopian communes (called Nashoba), salacious goings-on were apparently going on: black-white free love and corruption, among other things. The story hit the international headlines and ruined the future of the settlement. However, it did not ruin that of its temporarily absent founder, an upper-middle-class Scotswoman, Frances Wright, who then returned from London to the States, closed Nashoba, and took her message of miscegenation and mixed-race children-better-able-to-withstand-the-climate-of-the-South to mixed

Yearsley's other introduction to literacy came from Hannah More's pal, Elizabeth Montagu, for fifty years the hostess with the mostest at Montagu House, her latest-rave-architecture-and-décor pile in Central London, where on occasion she'd offer breakfast to seven hundred or more. Most of the time it was soirées to discuss literature with the society-page big names of the moment. Montagu also made her name as a bluestocking (the ladies wore stockings in shockingly indiscreet blue instead of respectable black silk). Possessor also of a zillion bucks, Montagu traveled a lot, wowed the Parisian demimonde, and took on protégés like serious ex-Freechurch preacher Richard Price. If your insurance recently paid up on the dot without a quibble, you have Price to thank. After extensive study of the birth and death registers, Price produced the kind of math that enabled the early insurance companies to relate premium to life expectancy, lower their charges, attract more clients, and make everybody as happy as anybody can ever be with insurance contributions. Price was also one of the loonies who supported the American independence movement. So indebted did the new United States feel toward him, in 1783 he got an honorary Yale degree when Washington did, and Congress asked him over to run financial matters. Alas, he was otherwise engaged, being at the time private secretary to Lord Shelburne, Prime Minister that year.

Long before, Shelburne had been in charge of administering the American colonies and was such a conciliator that, who knows, had he kept the job, the War of Independence might never have happened. But Shelburne explained himself to Parliament so badly (and, anyway, was always maundering on about how corrupt members of Parliament were) that he got the boot, and he was only back in power after the American thing was all over. Still, he did what he could to repair the damage, with a postwar peace treaty. His negotiator was Richard Oswald, whose wife's family owned property in the Caribbean and America and who spent months in France arguing with the canny B. Franklin and his diplomatic team. After ironing out various details, it all came down to a compromise: If the Yanks would drop claims to Nova Scotia and Canada, the Brits wouldn't insist on compensation for the Loyalists who'd lost everything in the war. Compromise was

audiences (white, male-female) all over the East Coast. Wright enjoyed a reputation as the "Great Red Harlot of Infidelity" and was shockingly well known for her numerous lovers, who included the Marquis de Lafayette. On her trip back to the States just after the Nashoba scandal broke, she was accompanied by Frances Trollope (mother of the future novelist) and her failure of a husband, on their way to set up a fancy-goods store in Cincinnati. When Mr. Trollope failed yet again, he and his wife traveled for a while. Three years later, when they finally returned to the United Kingdom, Frances Trollope decided to put her travel notes together in *Domestic Manners of the Americans,* an instant success and one which started her career as a writer—and saved her from starvation when Mr. Trollope inconveniently died in 1835.

Trollope went on to write more than twenty bread-winning novels, which then sank into total oblivion (samples: *Mrs. Matthews, or Family Mysteries; Second Love, or Beauty and Intellect).* Trollope also traveled around Europe and wrote about those trips. One example, about a visit to Paris, included a chapter on another woman of character, Amadine-Aurore-Lucille Dupin, a.k.a. George Sand. It's hard to remember with whom George Sand *didn't* sleep. After an early marriage with an older man, she successively had affairs with a lawyer, a law student (Jules Sandeau, who gave her the idea for her pen name), a famous novelist (Mérimée), a famous poet (Musset), a very famous composer (Chopin), and an unknown engraver. Not to mention those others (male and female) unrecorded but well known to Paris society at the time—as was Sand herself. Balzac, Flaubert, Delacroix, Hugo, de Vigny—all fell under her considerable spell. Sand wrote novels and plays about (guess what) love affairs. She also became attracted to socialism, out of which came a book that was translated in 1845 by an American writer, F. G. Shaw, for *Harbinger,* a reformist periodical published by George Ripley, former minister turned Transcendentalist, who founded Brook Farm. This was yet another utopian project where everybody was equal and self-sufficient, and produced oil lamps, teapots, shoes, nature books, and window blinds. Like all other utopias, it eventually folded, and Ripley took himself off to the *New York Tribune.*

In the four years *Harbinger* was in print, it also employed the writing talents of Stephen Pearl Andrews, who started his own commune: Modern Times, on Long Island. This commune also fell apart—free love again. Around 1860 Andrews came up with an idea for what he called Universology ("everything is linked to everything else") and invented a new universal language, Alwato, which everyone would speak. As far as is known, Alwato was spoken only by Andrews. In 1871 he got a job with the new *Woodhull & Clafin's Weekly,* described by its publishers as

reached, and everybody signed on November 30, 1782. And that was that—till 1812.

Meanwhile, a couple of years later Oswald died, and six years after that, so did his widow. This event then achieved literary immortality because on its way to her burial in Scotland, the funeral party happened to turn up at an inn one foul night and turfed out some common folk who otherwise might have had a warm bed. One of these, forced to ride twelve more miles through a snowstorm to the next inn, wrote a scathing verse about the event. And because he was Robert Burns, it got into print and the world got to hear about it. The whole Burns myth is still alive and well on "Burns Night" all over the world, when exiled Scotsmen scoff up a cooked form of sawdust known as haggis, and drink copious whisky toasts to the ploughman poet—which he wasn't, being book educated and speaking French, and with a job with Customs, posh literary friends in Edinburgh, and more affairs and illegitimate offspring than even Burns could remember. But for a' that and a' that (as he might have said), Burns wrote some haunting stuff. His "My luve is like a red, red rose" is a poem without equal in beauty of expression. Burns's mythical "ploughman poet" label originated with a hack named Henry McKenzie, who wrote several novels and plays, all of them failures, and had much the same luck with various essay magazines. It was in one of these, the short-lived *The Lounger,* that in 1786 the remark about "ploughman" Burns appeared. Two years later, Henry produced *Account of the German Theatre,* even though he spoke no German. Kicked off Scots interest in German thought and in particular the whole Romantic mishmash coming out of places like Jena.

But to crown an undistinguished career, after creating a myth, Henry McKenzie also courageously dispelled one. In 1805, ahead of most, he pronounced as fake the recent Romantic rave success, the epic poem *Ossian,* a purported third-century Celtic piece, purportedly discovered by James Macpherson. McKenzie also shot a lot, on occasion with the otherwise-discerning novelist Sir Walter Scott and science stars Humphry Davy and William Hyde Wollaston. In an era before doctorates, Wollaston was able to write authoritatively about pathology, physiology, chemistry, optics, mineralogy, crystallography, astronomy, electricity, mechanics, and botany. He also wrote a paper titled "On Fairy Rings" (1807, if you're interested). He appears to have been primar-

the "Organ of Universal Science . . . Government . . . Religion . . . Language." The publishers were already notorious. The delicious Victoria Clafin Woodhull and her lovely sister Tennessee had spent time as hookers, jailbirds, and spiritualist table rappers. In the course of this last, they managed to persuade the megarich financier Cornelius Vanderbilt that they'd put him in touch with his dead wife. So he bankrolled them to become the first female stockbrokers ever, and they became rich enough to found a radical magazine—and, in Victoria's case, in 1872 to become the first-ever female presidential candidate. The fact that women didn't have the vote didn't seem to deter her.

One of Woodhull's backers, the influential and well-known Rev. Henry Ward Beecher, pulled out after hearing the candidate's views on love and marriage, so Woodhull threatened to publish the rumors going around about Beecher's private life. Beecher sent his sidekick, Theodore Tilton, to shut her up. Woodhull promptly seduced Tilton, and he became her campaign biographer. Then it turned out Tilton was the other half of the Beecher rumors. He was being cuckolded by Beecher, at this point a famous abolitionist, nationally known preacher, brother of the author of *Uncle Tom's Cabin,* and writer of *The Life of Jesus Christ.* Also, it now appeared, adulterer. Woodhull went into print, Tilton took Beecher to court, and in 1875 the trial was a national cliffhanger. In the end, a hung jury (perhaps unable to face the social consequences of a guilty verdict) exonerated Beecher, the whole thing was swept under the carpet, and Victoria and Tennessee headed off for England and wealthy marriages.

Beecher's lawyer in the case was Benjamin Tracy, Civil War hero (Medal of Honor) and former federal attorney, who in 1889 became Secretary of the Navy under Benjamin Harrison and set about bringing American deepwater warships (obsolete wooden hulls, a few ironclads, only one-third ready for service) into the twentieth century, with a proposal for twenty armored battleships to be stationed on both oceans. By the time Tracy left office in 1893, the United States had nineteen steel ships and a high-tech navy ready to carry American influence overseas. Tracy forced through unheard-of reforms, such as a requirement for competitive tendering on the steel needed for guns and armor. When the armor was ready, Tracy demanded live-round, point-blank testing by both major domestic suppliers, Carnegie and Bethlehem Steel. Bethlehem changed its focus from rail manufacture to guns and armor for the contract, which was why, in 1892, Bethlehem was able to handle a civilian order for the largest single piece of steel ever forged in the United States. It was the 46.5-ton, 45-foot-long axle of the Ferris Wheel erected for the 1893 Chicago World's Fair. The aim of the exercise was to do better than the Eiffel Tower. The Ferris Wheel did that: 264 feet high and 825

ily interested in the very small, producing a microscopic amount of palladium, once he'd discovered it. He then invented the tiny reflecting goniometer, for measuring crystals. A French fan named the mineral wollastonite after him. Not much else to say. One of his pals was another generalist, Andrew Ure, who taught chemistry for a while in Glasgow and gave evening classes to craftsmen in his spare time. In 1818 Ure got Byron excited by applying electricity to a corpse and getting it to twitch. In 1821 he produced a *Dictionary of Chemistry*. He helped get modern chemical symbology accepted and, in the words of one eminent biographical dictionary, made "no significant scientific contribution."

One of Ure's Glasgow pupils was David Dale Owen, a marginally more exciting fellow, who went to United States with his dad Robert, the founder of another utopian commune, New Harmony, Indiana. He stayed in the States when Dad went home and became state geologist for Indiana, Kentucky, and Arizona (his brother R. D. was a local Democratic bigwig who later went on to Congress) and classified all the geology of interest in two hundred thousand square miles of the upper Mississippi Valley. He was also instrumental in giving the Smithsonian that dreadful castle bit (R. D. having shepherded through Congress the bill that accepted Smithson's bequest). As a geologist, Dale got letters from anybody digging holes. One of these was Hamilton Smith, coal-mine owner in Cannelton, Indiana, who had a dream of turning the place into a kind of industrial park. To this end, he was involved in a cotton mill and pushed hard for railroads to come into the area. Smith therefore wrote (an indefatigable scribbler, to judge by his list of correspondents) to Henry Varnum Poor. Poor was hot stuff on everything to do with railroads, having perspicaciously noticed that they probably had a future.

From 1849, with his *American Railroad Journal,* Poor became your guide to everything on rails. If you wanted statistics on the tiniest railroad-related detail, Poor had it. In 1868 Poor and his son started to publish an annual manual on the subject, and it became bedside reading for those with investments in anything that used the iron horse. So then Poor started an investors' information service. It would one day become the Standard & Poor's 500 that the world checks with trepidation every morning. One of Poor's younger contributors to the *ARJ* was high schooler Alex

TRACK ONE

feet in circumference, the Wheel carried sixty people in thirty-five cars.

Within a year, there was a wheel in London, built by engineer Walter Bassett. Bassett then hired another engineer, Cecil Booth, to do the same thing for Blackpool, Paris, and then one (built in 1896, destroyed in World War II, restored in 1945, starring in the classic movie *The Third Man,* and still working today) in Vienna. Booth then set up his own engineering firm and in 1901 patented his Puffing Billy. Horse-drawn, gasoline-powered, and the size of a large refrigerator, Bassett's gizmo sat outside the hotels and mansions of London (and, for the coronation of Edward VII, Westminster Abbey), connected to the buildings by giant hoses that went through the doors and windows, and sucked out the dirt. Puffing Billy was the world's first vacuum cleaner.

END TRACK ONE

Holley, who wrote a piece on cutlery (his father made it). Holley grew up to become a locomotive freak, visited Europe for a compare-and-contrast report on the railroads, and then became a technical writer for the *New York Times*. On another European fact-finding tour (into steelmaking and iron making), Holley visited Sheffield, England, and became convinced that steel was the future and iron wasn't. Almost single-handedly, he established, modified, and improved the 1857 Bessemer (blow air through molten metal) steelmaking process. The great 1875 Edgar Thomson plant, which turned Pittsburgh into Steeltown USA, was Holley's design.

Meanwhile, Bessemer himself was long out of things metal and into things maritime, with an idea for a circular steamship. The designs lay on the drawing board until the visit of a Russian Admiral called Popoff, who was attracted to the concept (flat bottom, shallow draft, armor-plating, all-round field of fire, highly maneuverable) and commissioned a Glasgow shipbuilder to build it. In 1868 one of the young naval architects working on the "Popoffka" ships was John Thornycroft, who then helped the British navy solve an awkward problem. The navy had the newly invented torpedo, which then turned out to be useless without a fast launch platform. By 1877 Thornycroft had built the torpedo boat *Lightning,* and went on to develop lighter and faster versions, using a new type of hull he had designed, which allowed a boat to skim the water, rather than cutting through it. By World War I, he was making high-powered coastal torpedo "scooters" and decoy ships that would ride over minefields. Thornycroft's long-term idea (patented in 1877 but never built) was to reduce drag on a ship by adding a concave hull, below which air could be trapped between hull and water. The theory was given added support thanks to the later discovery by airplane pilots that they got extra lift (when flying just above terrain) from the so-called ground effect.

END TRACK TWO

AND FINALLY . . .

In 1959, Christopher Cockerell put together a reverse version of Cecil Booth's Puffing Billy and the air-cushion theories of John Thornycroft to produce the Saunders-Roe N1—otherwise known as the hovercraft.

TWENTY

1673: SIEGE OF MAASTRICHT TO VENDING MACHINES

On June 10, 1673, Louis XIV of France, plus forty-five thousand soldiers, fifty-eight cannons, and enough food for six weeks, settled into position round the Dutch town of Maastricht for yet another one of those sieges. Back then, sitting it out was the usual way. Eventually the inhabitants would starve or the besiegers would run out of ammo and go away. This time, however, things were different thanks to the new French secret weapon: trench warfare. Dig a trench six hundred yards from (and parallel to) the fortification wall. Install cannons and commence firing to protect your sappers, who are now digging a diagonal trench toward the wall. When they get to within three hundred yards from the wall, they stop and dig another trench parallel to the walls. You install cannons and so on. Repeat until you are at the foot of the wall. Blow a hole in it. Send infantry through the hole. Three weeks after the siege of Maastricht started, thanks to trench warfare it was over.

Responsibility for all the spadework rested with Louis's right-hand engineer, Sebastien le Presle de Vauban, foot soldier, artilleryman, gunpowder maker and surveyor, who wrote about everything from taxation to trade, transport, shipbuilding, and pig farming, and was a general smartie well into old age (when he wrote *Various Thoughts of a Man Without Much to Do* and invented the socket bayonet). Vauban also predicted that the population of Canada in A.D. 2000 would be fifty-one million (on the day, it was only thirty-one million, but Vauban had been reckoning on crowds of French immigrants that never arrived after France lost Canada to the Brits). Vauban left his mark throughout France with the thirty-three star-shape fortresses he built and the three hundred he repaired. By 1703 he was marshal of France. Earlier, Louis asked him to take a look at the new Canal du Midi being built, from Toulouse (Atlantic) to Sète (Mediterranean), to save the trip round Spain. His survey, in 1684, noted a few required improvements, one of which was an aqueduct, which he built. When the Canal went operational in 1686, twelve thousand laborers had dug a ditch 175 miles long, 75 feet wide, 6 feet deep, and furnished it with 191 locks. It's still there, the joy of tourists.

In the seventeenth century, the canal was big-time toys-for-boys. One of these boys being the Englishman Francis Egerton, third duke of Bridgwater. Who as a child was such a boob that his family thought of disinheriting him for stupidity. After seeing the Canal du Midi (during his obligatory culture-vulture Grand Tour of European arts and crafts), Francis went home and for the rest of his life did little else but dig. He lived for canals, partly because most of what he had inherited was coal mines and the buyers in Manchester were a long way away by rutted and expensively slow country road. A canal would speed the coal and cut the costs. So Egerton consulted a local think tank, started excavating, and in 1761 the first boatload of his coal floated to market, to be sold at half the previous price.

That local think tank (known as the Lunar Society because they met at every full moon so as to have bright nights for riding home) was made up of a bunch of scientifically inclined amateurs and included another rapid-transit freak, Josiah Wedgwood. His interest in canals sprang from needing easier ways to carry raw material (clay) and fuel (coal) to, and finished pottery from, his custom-built factory and workers' village, Etruria. Wedgwood was an industrial big cheese, being master potter to the queen and who by this time was already known for the fancy china of which you may well have some pieces. Among the other things Wedgwood made were medallions, one of which (1765) portrayed yet another member of the think tank. Exper-

TRACK TWO

On the night of June 21, 1673, during the Maastricht siege, a middle-aged lieutenant of musketeers in Louis XIV's army was fatally shot in the throat. Today few people remember much about Louis or the siege, but everybody who reads novels, goes to the movies, or watches TV knows who d'Artagnan was. Except he wasn't. His real name was Charles de Batz, and he never met the other "three musketeers," or the "evil Richelieu" (who was sixteen years dead when de Batz became a musketeer in 1658). In 1661 his career had been boosted when he was ordered to arrest the previously powerful Minister of Finance, Nicholas Fouquet.

Fouquet had started well (born rich, lending Louis XIII tons of money, fighting on the king's side against rebellious nobles, and becoming Finance Minister), then used his position, as they all did, to close a few less-than-kosher deals and make himself even better heeled. Then things went really wrong. Louis XIV's favorite adviser, Colbert, dished the dirt on Fouquet, so when, in 1661, Fouquet invited Louis to his amazing new château of Vaux le Vicomte for an extravagant, let-me-explain dinner party, complete with lavish theatrical entertainment, the king saw a rich-and-famous lifestyle better than his own and had Fouquet slapped in jail. Still, some good came out of it. Fouquet's relative, Pierre Beauchamp, had arranged the evening's entertainment and went on to become Louis's dance master, then superintendent of the king's ballets, and then, in 1671, director of the Royal Academy of Dance. Along the way, Beauchamp invented choreography and the basic steps of modern ballet. In 1706 came an English translation of Beauchamp's book by choreographer John Weaver, about to burst on the scene at Drury Lane Theater with his "danced dramas" that for the first time included dancers pantomiming emotions with gestures and movements. In 1708 Weaver had the good luck to get support from a couple of influential scribblers, Joseph Addison and Richard Steele, who started the *Tatler* and the *Spectator*, magazines for readers who wanted to know what they ought to be talking about in the fashionable coffeehouses of the capital—with a political slant, of course. Journalism back then was more party propaganda than it was anything else. In return for these services, writers were given salaried "jobs" in obscure corners of the government's bureaucratic machine.

imenter and libertarian Joseph Priestley (to whom Wedgwood gave money and lab equipment) was a short-statured Unitarian preacher with a stammer, who predicted that the Second Coming would happen at the latest by 1814. Priestley failed miserably in the pulpit, taught for a while, then spent most of his spare time noodling around with gases. Easy stuff to do when you had a wealthy wife (sister of major ironmasters) and lived next to a brewery where the vats gave off a lot of carbon dioxide. In 1775 Priestley discovered oxygen, then invented soda water, wrote about electricity, and became a science superstar with friends like Franklin and Lavoisier. And enemies like the 1791 mob that burned down his lab and seven other houses around it after Priestley started publishing how much he approved of the American rebels and how much more he approved of the French revolutionaries. Thus deterred, in 1794 Priestley left for friendlier shores (a.k.a. Pennsylvania).

There he raised the hackles of another Brit whose nickname was "porcupine." William Cobbett (teaching English to Philadelphia French immigrants) made his name with a pamphlet lambasting Priestley's rapturous reception in the States. Cobbett was that most unusual of types, a continuing Loyalist in the postwar, newly independent United States. For several years (in Philadelphia and then in New York), he compounded the felony by haranguing Republicans. Back in the United Kingdom (to a hero's welcome), his aggressive reformist stance soon alienated the authorities and he spent time in jail, then spent time back in the States, and finally took a tour on horseback throughout England to research one of the most beautifully observed descriptions of contemporary agricultural life, titled *Rural Rides.* (Well worth the read.) On occasion, Cobbett appeared to take sides just for the fun of it. For example, in the autumn of 1808, he surprisingly supported the management side through two months of the famous "Old Prices" riots. The rioters weren't, as you might expect, workers but patrons of the newly reopened (after the recent fire) Covent Garden Theater.

The management in this case included the greatest actor of the time, John Kemble, who, with his sister, Mrs. Siddons (the greatest actress of the time), had been laying 'em in the aisles for nearly twenty years and making enough money to buy a share in Covent Garden. The fire ruined him, and it was only by borrowing heavily that he was able to rebuild. And he had to raise the prices to help pay for the costs—hence, the riots. When the dust had settled, Kemble went on with his career, in due course retired, and sold his massive collection of plays to the sixth duke of Devonshire, a gent with a literary turn of mind. William Cavendish (the duke) led a quietly elegant life, collecting art and landscaping his many acres. At one point, this

Once a week, Addison and Steele would meet and greet their political bosses at the famous Kit-Kat Club, where wine and gossip flowed freely. The club was the idea of publisher Jacob Tonson, whose trick was to invite promising young writers and sign them up cheap. Traditionally the Kit-Kat Club chose a beautiful woman as "toast of the year," and her proposer etched a poem to her on the club wineglasses. Back in 1697, the toast had been the precocious eight-year-old, Mary Pierpoint. Pierpoint later became Lady Mary Wortley Montagu, when she married a dumb aristocratic ambassador, with whom, in 1716, she went to Istanbul—and wrote a marvelous collection of letters about life in Turkey. In 1718 Lady Montagu brought back the secret of inoculation against smallpox. Apart from spreading the word on the new technique, she also spent time writing witty sallies and becoming very close pals with (and then enemy of) the wasp-tongued Alexander Pope. She became even closer pals with half-her-age androgynous Italian Francesco Algarotti, in London to gather material for his *Newtonianism for Ladies.* In 1739 Montagu suggested they "meet" in Italy and left for the assignation. Algarotti never showed, heading instead for a better position (court chamberlain and very close pal) with Frederick the Great of Prussia. On his way to London, Algarotti had stopped in Paris and accepted an invitation from the beautiful Émilie du Châtelet to come visit at her château in Champagne, where she and lover Voltaire were busy (with Emilie's absent military husband's approval), each writing a book on Newton. Du Châtelet's writing program was pretty ambitious: a piece on the nature of fire, a book on the rational nature of language, elements of Newtonian philosophy, and finally nothing less than a translation of Newton's *Principia Mathematica.* But even du Châtelet needed help in prepping. In 1739 she hired a young algebra tutor, Samuel Koenig, who'd studied under Swiss math mavens the Bernoulli brothers. Koenig wrote something on the structure of honeycombs and was later taken under the wing of French geodesist and general science whiz, Pierre de Maupertuis. Maupertuis arranged for Koenig to get a job and to be elected to prestige membership of the Prussian Academy of Sciences, which Koenig repaid in 1751 with an article in which he accused Maupertuis of plagiarism. A very ugly situation developed (Koenig couldn't prove the accusation because an essential document was in the hands of a Swiss who had been

exercise involved moving an entire village because it spoiled his view. And at another point (1844), he erected the biggest gravity-fed fountain in the world next to his ducal pad, Chatsworth House. The plan was to get the waters spouting in time for the visit of the Czar of Russia. After six months of night-and-day endeavor, involving the construction of a massive reservoir 350 feet above the house, the "Emperor" fountain finally shot its waters 296 feet into the air, flabbergasting all—except the czar, who never turned up.

The man responsible for this unobserved wonder was the duke's estate manager, Joe Paxton, who'd met the duke one day in 1826 at a time when Paxton was considering emigration to the United States, fed up with tree grubbing for the horticultural society (the duke was the president). Liking the cut of Paxton's jib, the duke hired him. By 1836 Paxton was building the world's largest conservatory, at Chatsworth. A year later, he succeeded in growing the world's biggest lily (twenty feet across), the recently discovered and utterly excessive *Victoria Regia* (so-named after the new queen, and today known as *Victoria amazonica).* Paxton then went on to fame and fortune (as Sir Joseph) when he designed (on a piece of blotting paper) the main building for the 1851 Crystal Palace Universal Exhibition and got the job over 233 other contenders. This 300-foot-long, 145-foot-wide, 60-foot-high all-glass object of general amazement was held up by a pattern of girders, which Paxton said had been inspired by the supporting power of the underleaf ribbed structure of *Victoria Regia* (whose strength he had demonstrated by standing his twelve-year-old daughter on the lily pad).

The wonder plant had first been discovered in fittingly imperial fashion by a German explorer, Robert Schomborgk (although four other explorers later claimed the fame), who in 1831 had been asked to go and map British Guiana, the only British possession on the South American mainland. In 1840 he was sent back there again, to make sure everybody knew where the British Guiana-Venezuela border was (it became known as the Schomborgk Line), so that the next-door Venezuelans could end those ludicrous claims about owning parts of what was indubitably British Imperial property (especially after gold was discovered there in 1850). Britain sniffily ignored the Venezuelans (they were, after all, only Venezuelans) until they pulled a fast one and brought the United States in on the act, appealing to the Monroe Doctrine about foreign powers keeping out of the American continent. In 1895 the immensely harrumph Sir Lyon Playfair sat (as British representative, together with some Americans and Germans) on the Border Commission that finally said the Brits were in the right.

Playfair was a failed chemist and successful committeeman, who

recently decapitated) and lasted the rest of Maupertuis's life.

Maupertuis did too many things to list. Mainly, he went to Lapland and proved the Earth was a flattened sphere (key for accurate navigation), wrote a major work on heredity (based on study of people with extra fingers), and tried to develop a mathematical way of measuring happiness. One of his pals was the only man whose first name was that of the round church on whose steps his foundling form had been originally discovered (Jean le Rond d'Alembert). In an era of generalists, d'Alembert was as general as you get: math, religion, music, law, politics, philosophy, Latin translations, and science editorship of the *Encyclopédie* (the ultimate expression of the Enlightenment), in which he set the style for science with his "what you see is what you get" perception-based view of the world. Lucky d'Alembert lived with the beautiful Julie de Lespinasse, hostess at one of the better salons in Paris. Another chatter place he visited was run by Madame Necker (wife and personal financial adviser to the French Minister of Finance), whose daughter Germaine de Staël ran the most famous salon of all. De Staël's trick was to receive her guests in bed, dressed in something that displayed her to advantage. The purpose of salons was to talk about everything, which suited de Staël fine. There was nothing she didn't have an opinion on, including what was wrong with Napoleon, which was why at one point she had to slip away to Germany, where she talked the local eggheads to death (as Schiller noted) and bored Goethe, while dashing off her personal description of the local Germanic culture.

In 1804, while in Rome, de Staël met a budding young American writer, Washington Irving, doing a two-year tour for health reasons (also to develop his talents) before returning to certain literary success in New York, where European experiences would boost his stock. They did. In 1807 he started *Salmagundi,* for a year or so the must-read magazine of New York sophisticates (who found themselves caricatured in it, along with various satirical pieces on the latest fashions literary, political, and otherwise). Irving went on to fame and fortune with *Rip Van Winkle* and *The Legend of Sleepy Hollow* as well as a diplomacy-literature career that spread the word on the Far West (Ohio) and made him the first internationally known American writer. Such was not the case for his erstwhile colleague on *Salmagundi,*

oiled his way into the good graces of the powerful and sat on anything that was worth sitting on. He was chief adviser to Prince Albert for the Crystal Palace Exhibition and also appeared on government commissions set up to study herring fishing, compulsory education, urban hygiene, and the Irish potato famine. Last, but not least, in 1870 he presented a petition, signed by most of the country's great and good, for a halfpenny postcard stamp rate. A few months later, Parliament passed the necessary act, and the postcard craze really took off. In response, in 1883 Percival Everett invented a machine that would give you two postcards if you put a penny in it.

END TRACK ONE

James K. Paulding. At the time, Paulding was making the grade with sketch pieces, travel shorts, and reviews on everything from social criticism to the latest craze for phrenology (bumps of knowledge). Paulding's novels described life on the frontier, and in 1830 he wrote a play *(The Lion of the West)* whose character Nimrod Wildfire was based on, and made the reputation of, Davy Crockett, as the backwoods hell-raiser who could ride his pet alligator up Niagara Falls.

Crockett fostered his image as the simple country boy, hunter, and marksman, and got elected to Congress, where he stayed until 1835 (with one hiatus) for three terms. When he was at last defeated, he headed for Texas and what he hoped would be a political comeback. Alas, it was not to be. At the Mission San Antonio de Valero (a.k.a. the Alamo) on March 6, 1838, when two thousand Mexicans overran the 182 defenders, Crockett was taken prisoner, bayoneted, and then shot on the orders of the Mexican general López de Santa Anna. At the ensuing battle of San Jacinto, Santa Anna was captured and returned to Mexico. By the time he went into exile, he had been president of Mexico five times (once for life) and had run the country, one way or another, eleven times.

In 1865 he found himself in Sailors' Snug Harbor, Staten Island, hoping (and failing) to drum up support for a triumphal return to power. But before moving to Cuba, and a few final years of obscurity in Mexico, Santa Anna left one other mark on the world besides the state of Texas, whose existence he had helped to ensure, thanks to events at the Alamo. He introduced Staten Island photographer Thomas Adams to chicle, the tree resin Mexicans chewed. Adams flavored it, shaped it into small balls, and turned it into chewing gum.

END TRACK TWO

AND FINALLY . . .

In 1888 Thomas Adams used Percival Everett's slot-machine idea to sell his Tutti-Frutti gum in vending machines on the elevated stations of the New York subway.

TWENTY-ONE

1786: *THE MARRIAGE OF FIGARO* TO STEALTH FIGHTER

The first night of Mozart's *The Marriage of Figaro,* in May 1786, at the National Theater in Vienna, got mixed reviews. Some of the audience loved the revolutionary tone: women playing a key role in the plot, a story about social conflict, servants as good as their masters, bourgeois good sense versus aristocratic silliness, a generally reformist feel instead of the usual vacuous drivel. Even the Emperor liked it (he'd already had the more radical bits watered down).

The French play of the same name (from which Mozart had snitched the idea for *Figaro)* caused a public sensation long before it was staged. For two years, while Louis XVI's censor tried to have the text "modified," the rumors and gossip spread. On opening night, there were five thousand fans waiting outside the box office, it was a sell-out standing-room-only performance, and the unheard-of sixty-eight-night run made the author, Caron de Beaumarchais, very rich. It was just as well, because he'd lost a fortune covertly bankrolling the American War of Independence. Trusting government promises from both sides, he personally paid for enough ammunition, gunpowder, and guns to equip twenty-five thousand American rebels. In the end, the United States and France never settled Beaumarchais's bill. He stayed solvent only by marrying two rich widows (one after the other).

Early in his career, Beaumarchais advised Jacques Necker, who was on various occasions appointed French Director of Finances. Necker, a rich banker who lent the government money in 1777, promised to get the French economy out of a hole (with the original: "Watch my lips: no new taxes"), published his budget (a great work of fiction), was fired twice, and on the second occasion left things in such a mess the result was the French Revolution. One of his pals put it thus: "Necker never understood economics."

The pal was a one-legged American politico named Gouverneur Morris, who wrote the opening words of the U.S. Constitution: "We, the people of the United States." ("United" was his idea for keeping the former colonies together.) Morris also came up with decimal coinage, lobbied for the Erie Canal, proposed that the president be elected for life, and became U.S. minister to France. There he became involved in the questionable behavior of Silas Deane with regard to French financial support of the war. Deane was American commissioner in France, setting up intergovernmental postwar trade agreements, and it was he who claimed that the war chest had been a personal loan from Beaumarchais and not the French government. His fellow commissioner, Arthur Lee, disagreed and accused Deane of everything from insider dealing to money laundering. In the end, Deane lost the argument and most of the money he might well have been laundering.

It was Deane who in 1776 had recruited the nineteen-year-old Lafayette to serve as a major general in the American army. Once again, Deane was up to his usual tricks. The commission turned out to be invalid, and Lafayette had to put up with unpaid volunteer work on Washington's staff. He then had a good war, got his promised pro-

TRACK TWO

Mozart's last great opera (five years after *Figaro)* was *The Magic Flute.* The plot included an initiation, three veiled ladies, three boys, a fire-and-water trial, three temples, a padlocked mouth, a serpent, a High Priest, and a pyramid—little more than operatic fun and games to you and me but deeply meaningful to a Freemason. Mozart became a master Mason in 1785 (in the Vienna "Charity" Lodge) along with his pal Haydn and half the nobs in Austria. Chances are he also met the grand master of the Vienna Lodge of True Harmony, Ignaz von Born, during the regular music-and-chatter soirées they both frequented at Countess Thun's place, where she tinkled the ivories before graciously letting Mozart have a go. There are those who say that Born and Freemasonry inspired *The Magic Flute.* Born was a mineralogist and his lodge was keen on social reform and science. This may be why he also managed to recruit a naturalist by the name of George Forster.

Forster's family had started in England, emigrated to Germany, and then returned. By Born's time, Forster was back in Germany again, having left London in a huff after a run-in with the navy: Forster had sailed together with his dad on Captain Cook's voyage, and the deal had been that the father would write up the trip. In any event, Cook said he wanted to be the author, so that was that—except for Forster, who, not bound by any agreement, in 1778 came out with *A Voyage round the World,* beating Cook to the post and annoying the powers-that-were. With suddenly dimmer prospects, Forster lit out for a better life in foreign parts. He ended up traveling down the Rhine with none other than the great about-to-be-explorer Alexander von Humboldt, who doubtless got how-to tips from Forster on travel-book writing, before Humboldt then went off to South America for five years to get material for his own. By 1828 Humboldt was such a science whiz that he opened the Berlin Science Conference, before a host of international stars such as computing ace Charles Babbage, electromagnetics maven Hans Christian Oersted, and a fat, hypochondriac, self-taught chemist from Sweden named Jons Berzelius.

Who's the guy you blame if terms like $C_8H_{10}N_4O_2$ make your brain turn to porridge. Modern chemical symbology belongs to Berzelius—so do (he discovered them) cerium, thorium, and selenium. Berzelius was also the guru of the blow-

motion, and became a Great American Hero. Right up to 1825, he would return to the States and be cheered wherever he went.

Back home, he became a moderate radical and survived the French Revolution. In 1789 he was Louis XVI's jailer when the king tried his famous escape, during which Lafayette nearly came to a sticky end, when the mob turned ugly. Lafayette's aide-de-camp, Alexandre d'Arblay, spent the rest of his life dodging back and forth from England to France. As a moderate constitutionalist, when Napoleon's fortunes waxed, d'Arblay's waned, and vice versa. During a wane, in 1793 d'Arblay met a young Englishwoman, Fanny Burney, and they fell in love. At the time, d'Arblay was trying (and failing) to raise a troop of exiled French cavalry, in anticipation of a Napoleonic invasion of England, and Burney wrote him a note saying, in effect, I can let you have a hundred pounds sterling and me. They married.

Burney had published two best-sellers, *Evelina* and *Cecilia* (the last line of which inspired Jane Austen's title *Pride and Prejudice),* and was the darling of the literary criticism crowd (including painter Joshua Reynolds, who had to be fed by a servant so he could read *Evelina* without his eye leaving the page). None of this helped the shaky d'Arblay financial situation. Another book made enough to buy a cottage, but a play failed on its first night at Drury Lane. Not even the great Sarah Siddons could breathe life into it.

The greatest actress of her time (some say the greatest ever) was born into a family of traveling players. The actor-manager David Garrick auditioned her in 1775, and after a shaky start, she was a box-office smash-hit for the next twenty-seven years. Siddons played every role in the canon better than anybody else, but her Lady Macbeth had them cheering. The public would have adored her even if they had known the fact that Siddons paid her sister to stay more than 150 miles from London. At one point, the sister had tried to poison herself (in Westminster Abbey) and clearly needed her head examined, which would have been no problem for the man who married Siddons's daughter. George Combe was the premier exponent in Britain and America of the wacky but rave-craze pseudoscience of phrenology. Phrenologists, starting with Gall and Spurzheim, the Austrians who invented it, believed the brain contained different organs that were responsible for different characteristics (if you had a swelling behind your left ear, for instance, you were a good lover). And the size of these organs could be measured from the protuberances on the skull caused by their interior growth ("bump of knowledge" starts with phrenology). This craniometric stuff was a boon to social reformers, who could now "measure" the characteristics of criminals and the poor and find out who needed "treatment."

pipe (blow air down a tube and through a flame to turn it into a fine jet burning at 1500° C), which he used to break down and analyze everything from pieces of meteorite and ancient Egyptian mortar to a Canadian fur trapper's gastric juices. Speaking of which, Berzelius spent forty years in nearly constant gastronomy (in Paris, forty-one banquets in as many days) and meeting rare birds of all kinds, including, in 1845, the "Swedish nightingale," the stellar coloratura soprano Jenny Lind, on her way to becoming the toast of operagoers from Stockholm to ninety-three cities in the United States. There, on a year-long tour in 1850, arranged by Prince-of-Humbug showbiz genius P. T. Barnum, she caused riots wherever she appeared and made enough money to stop appearing. Her admirers named things after her: streets, schools, a dam, a suite in the Willard Hotel, a clipper ship *(Nightingale),* and a cheese-based soup.

Lind also inspired a story by Hans Christian Andersen ("Nightingale"), written a few months after he met her during her visit to Copenhagen in 1843, when he became totally smitten. Alas, she felt only sisterly toward him, perhaps understanding what Andersen wouldn't admit till late in life (when he fell for the Hereditary Grand Duke Carl Alexander of Saxe-Weimar-Eisenach). Apart from this minor glitch, Andersen's life was like his "Ugly Duckling" story: After a lonely childhood, the son of a country shoemaker makes good, goes to university, and becomes a world-famous author, traveling all over Europe and reading his stories to enraptured audiences. But he did not read to Queen Victoria, whose invitation to meet he declined, on the grounds that the dress code would cost too much. Andersen's first international success was in Germany. His second, in 1846, was in England, thanks to translator Charles Bonar, living in Germany with Prince Thurn und Taxis, writing his magnum opus *Transylvania: Its Products and People,* and about to become foreign correspondent for the *New York Tribune.* Bonar started in England, back in 1831, as tutor to the children of John Constable, whom he helped with lectures on how to paint.

At the time, Constable was unknown in England (he remained so all his life, while painting dozens of masterly English landscapes), because the most Constable wanted was to live in bucolic nowhere and paint it. The English, therefore, knew nothing about the utterly boffo effect Constable had on French art, in 1824, when his *Hay Wain* was

Combe's prematrimony check of Siddons's daughter established that she had a large bump of benevolence.

In 1844, when he was in Rome, Combe met a well-heeled honeymoon couple from the United States, one of whom was the teacher Samuel Howe, on his way to tell the pope about his famous deaf pupil Laura Bridgeman. Combe and Howe promptly went off together to the galleries to measure the heads of the statues. Howe's bride was accustomed to being left behind like that, because Howe resented her family money, and was often absent doing good and being a presidential adviser. It was only after he died that the highly intelligent and talented writer he had married, Julia Ward Howe, was able to leave the house and come out of her shell. No surprise that she then founded the Association for the Advancement of Women, tried to set up a worldwide feminist peace movement, and wrote "The Battle Hymn of the Republic." In her later years, Howe was a frequent dinner guest at the home of Alexander Agassiz, rich (successful copper mine in Michigan's Upper Peninsula) and stimulating director of the Comparative Zoology Museum at Harvard—and a freak for starfish, sand dollars, sea urchins, and related bottom dwellers. President of the National Academy of Science and guru of the marine biology world, Agassiz designed nets to pull up specimens from specific depths.

Special nets were key to the work of a fellow oceanographer. Victor Hensen, German physiologist and amateur plankton lover, sieved these teeny creatures in his special mesh, then multiplied his catch by the number you first thought of, and (because plankton is the basis of the oceanic food chain) got a very approximate idea of how rich the catch would be thereabouts. This would be helpful for the German fishing fleets, he thought. Ernst Haeckel (himself big on sponges and jellyfish) didn't think so, and argued Hensen's little critters weren't evenly distributed enough to tell you meaningful things about their environment. Ecology (he invented the word) was Haeckel's thing, ever since his instant conversion to Darwin. It was Haeckel's development of a league table of organisms (all the way up from simple slime to complex human) that would, later on, make him the fave rave of the Nazis. Haeckel came up with a biogenetic law that "proved" how superior Europeans were to "savages."

He set up a Monist League (monad = primal-slime thing) to help spread the word, and in 1910 invited a highly respected chemist, Wilhelm Ostwalt, to become Monist president. Ostwalt believed in a world based on science, international peace, and a world language (he invented one named Ino). Ostwalt's specialty was catalysis, the process which speeds up chemical reactions, for which he got a Nobel in 1909. While working on how chemicals work together ("affinity"),

shown at the Paris Salon. The French artist Eugene Delacroix was so affected by Constable's use of color, he reworked his own chef d'oeuvre, *Massacre at Chios* (exhibited at the same show), four days later. Delacroix's first painting had been shown only two years earlier and had caused such a sensation the government bought it. From then on, Delacroix had a love-hate relationship with both government and the public. Either you loved his stuff because it was wildly, over-the-top Romantic, or you hated it for the same reason. Delacroix painted an extraordinary variety of topics: North African scenes, wild beasts, giant historical canvases, and (now and again) portraits of friends. In the 1840s, to get away from it all, he spent time in the country with Chopin, whom he regarded as the "truest of artists."

The feeling was not reciprocated. Chopin was a member of the I-hate-Delacroix school of thought. The painter's amour propre was saved by only the fact that one of his devoted fans was Chopin's live-in lover, George Sand (real name: Amandine-Aurore-Lucille Dudevant). Chopin was Sand's nth affair, an activity she specialized in when not writing highly successful novels and plays or getting involved in revolutionary politics and suffragetism. The eleven years Chopin enjoyed with her was a record, and in 1847 she inevitably left him for (many) others. Her feminist efforts were much admired by Elizabeth Barrett Browning, who'd written a couple of poems about Sand (before they met in Paris in 1852) while shuttling back and forth between England and Florence with her husband (and lesser poet) Robert. For years, before running away with Robert, Elizabeth had written poetry in a darkened room and had suffered from an obscure illness that left her weak and feeble. Robert persuaded her to emerge into the open air, take exercise, and eat healthy food. She recovered, and the two of them spent a life writing what some people regard as literary sugar.

Theirs was unlike the work of their pal Alfred Tennyson, poet laureate from 1841, and the archetypal harrumph Victorian. Even Queen Victoria loved Tennyson's verse: manly but not rough, sensitive but not soppy, and above all, British to the core, especially "The Charge of the Light Brigade" and a lot of brilliance about King Arthur. You felt there was no nonsense to the man, know what I mean? And yet, as with all Victorians, that was only half the picture.

TRACK ONE

he decided that the world ought to be seen in terms of energy, not matter, and invented the science of energetics. This idea went over like a lead balloon with his colleagues, especially the great Ludwig Boltzman, with whom Ostwalt had an almighty row at a conference. This was observed by a young mathematician, Arnold Sommerfeld, also there to explain (to those very few who would understand a word of it) his advanced theoretical-physics work in radiative scattering and such. Sommerfeld used math to describe the ways in which electromagnetic radiation bounced when it hit something. His lab, for instance, produced the first example of this by bouncing X-rays off crystals. How much the radiation would bounce, he determined, would depend to an extent on the shape of the object. Useless noodler's gobbledygook? Not, as would turn out, in combat.

END TRACK ONE

TRACK TWO

The dark, secret side to Tennyson was that he liked to be in curtained rooms, talking to spirits. Holding hands with serious scientists like Sir William Crookes, inventor of the radiometer and, in 1878, the Crookes tube (low-pressure gas in tube, electrodes, current, green glow, and mysterious rays). Soon everybody was playing with it, and in 1895 a guy in Wurzburg (named Roentgen) stuck his hand in front of the rays (he called them "X" for "unknown") and saw his bones. A year later, a Frenchman, Henri Becquerel, looking to see if glowing materials would produce other such rays, put some uranium sulfate next to a photographic plate in a dark cupboard, and later found the plate as if exposed. He told his graduate student, Marie Curie, about this, and the rest is radioactive history. Becquerel had heard about X-rays from his pal, mathematician Poincaré, whose work is so impenetrable I'll mention only that in 1904 he became vice president of the Astronomical Union.

The Union's president was American astronomer George Hale, who discovered that sunspots are cooler than the surrounding solar surface and that the sun is a magnet, and who laid the groundwork for the discovery of solar wind. He also planned the Huntingdon Library, fathered the discipline of astrophysics, and set up the California Institute of Technology with the help of his old teacher, chemist Arthur Noyes. Noyes had established his reputation at M.I.T., where he knew more about the physical properties of aqueous solutions than anybody, and in 1903 he taught a young visiting Japanese scientist, Yagoro Kato, who would go back to Tokyo and in 1935 set up one of the great Japanese electronics industrial giants, TDK. This, after he'd invented a new kind of "soft" ferrite material that made possible magnetic recording of sound, data, and video. It would be useful in combat.

END TRACK TWO

AND FINALLY . . .

In 1975 an American engineer, Bill Schroeder, used Sommerfeld's math to make thin, diamond-shape panels that would bounce radiation away in almost any direction but back to the radiation transmitter. This made the panels virtually invisible to radar. The radar signal had also already been partially absorbed by the paint on the panels. The paint was made of a binder containing particles of Kato's soft ferrite. These two components gave the stealth fighter, the F-117A, its strange, angular shape and made it undetectable.

TWENTY-TWO

1780: EDINBURGH OYSTER CLUB TO DNA

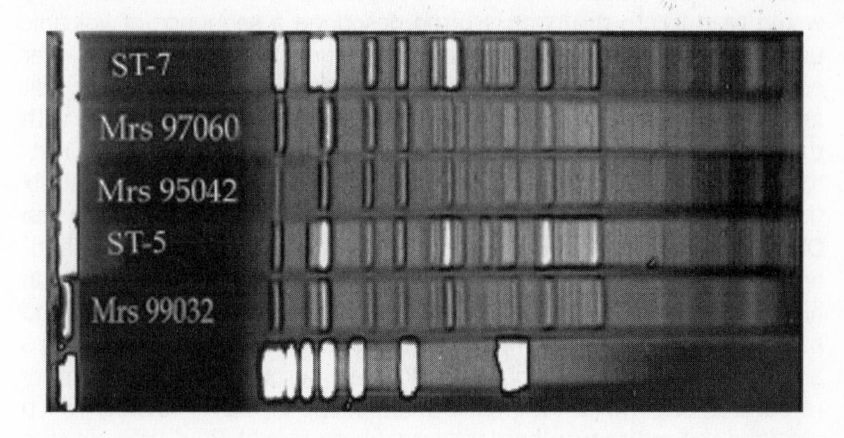

In the 1780s Edinburgh, Scotland, became known as the Athens of the North, when what became known as the Scottish Enlightenment flowered, funded by the profits pouring in ever since the recent union with England had opened up direct American markets for Scots entrepreneurs and Scotch whisky. At weekly dinners in the Edinburgh Oyster Club, the city's intellectual elite met, to sharpen their wits and enjoy founder-member Adam Smith's latest bon mot on the division of labor.

One Oyster Club founder was James Hutton, the man who, more than any other, made people aware there might be such as thing as evolution (the idea didn't later spring, fully formed, from the brow of Darwin). After failing to find work as a doctor in Edinburgh, Hutton had turned to farming his inherited acres, and any time he got the chance, he traveled in England, Belgium, and Holland looking at the latest agricultural practices and becoming increasingly aware (in a way most farmers aren't) of what lay under his feet. Hutton became convinced that the process through which rocks were eroded by the weather was probably nothing new and was probably something ancient. In 1795 these crazy-for-the-time musings took the form of a two-volume *Theory of the Earth* (so turgid and incomprehensibly written that it needed ruthless editing by a friend before anybody would read it). In the book, Hutton described a sequence of volcanic upthrust, followed by erosion to dust, which then sedimented over eons into layers of rock, followed by volcanic upthrust—then repeat as before. Processes, in other words, that were visibly uniform with those going on in Hutton's back yard. Uniformitarianism brought Hutton to the theologically shocking idea that the Earth wasn't about six thousand years old but nearer five hundred million. So old that when Darwin (and Wallace) got round to thinking about lengthy evolutionary processes, there was Hutton's geology to back them up. Early in his geologic wanderings, back in the 1780s, Hutton's illustrator and pal John Clerk had accompanied him on several rock-tapping trips around Scotland. Landlubber Clerk then surprised all and sundry (around 1782) with a little piece titled *Essay on Naval Tactics,* which changed the entire way battles at sea were fought (and first put into practice maybe by Rodney but definitely by Nelson). Clerk's thoughts make perfect sense but not to the old bores of the we've-always-done-it this-way Royal Navy, which for decades had played it by the book: line up your ships opposite those of the enemy, and then have each ship attack only its opposite number. And if you can't do that, go away. Clerk's weird new idea was that if the enemy line showed any weakness, you should clobber that spot with all the ships you could find and break the enemy line apart, then mop up. That is just what Admiral Rodney did to the French fleet in 1782, off Dominica. And because Rodney had an annotated copy of Clerk's essay, there's been a row ever since about whether or not Clerk gave him the idea.

Rodney's opposite number French admiral (captured in the battle) was a man well known to all Americans, since without his fleet in the Chesapeake Bay a year earlier, British general Cornwallis wouldn't have had his supply lines cut and his troops shelled by the Frenchman

TRACK TWO

Dr. Joseph Black can't have been much fun during the dinner-table talk at the Oyster Club, given his habitual diet (bread, prunes, and watered milk) and his tendency to spit blood. Not surprisingly, his doctoral thesis had been on indigestion. During the course of this dyspeptic research, he also established quantitative analysis in chemistry, and then, once in teaching mode, trained almost every future chemist of importance. As a Scot, Black was aware of the vital importance of whisky and how best (and most profitably) to distill it. Investigation of such matters led him to discover that steam was hot because it stored the giant amount of heat used in making steam from water. This discovery in turn helped Black's protégé James Watt to make a more efficient steam engine. It was also Black who lent Watt the money to do so and change the world. So let's hear it for coughing wimps on a diet.

Black's great pal and mentor was another medic-cum-chemist named William Cullen. He and Black played tag with chairs of chemistry at Glasgow and Edinburgh (one followed the other). Cullen also taught medicine, and produced the Great Nosology (or "naming") Book. The idea was that if you named all diseases and subdiseases and sub-subdiseases, and their symptoms and subsymptoms and sub-subsymptoms, you'd have a tool with which to diagnose and cure. Cullen's book included the immortal condition-and-symptom beloved of all medical students: "Nostalgia: a powerful desire to go home."

Both Black and Cullen taught and inspired a young visiting American, Benjamin Rush, who was in Edinburgh to get a medical degree (thesis, predictably: *On the Digestion of Food in the Stomach)* so that he could go back to Philadelphia and immediately be made professor of medicine (in the country of the blind the one-eyed man being king, so to speak). Working in Philadelphia and having met Franklin in London, it should be no surprise that Rush also became involved in events revolutionary, in the course of which he joined a group of similarly independence-minded types and eventually signed The Declaration. Sensing that revolution wasn't going to be easy and that the public needed to have their morale stiffened for the fight ahead, Rush suggested to a fellow dissident—a British expatriate, ex-corset-making scribbler named Thomas Paine, who'd left England under

in question (Admiral De Grasse), Yorktown wouldn't have fallen, and there might not have been a United States of Anything. That De Grasse was present at all, that day, was ironically due to the fact that the British fleet (which should have wiped him out a few days before) played it by the book and left the scene because they couldn't engage in the traditional one-on-one manner (see above). Poor old De Grasse spent most of his family fortune on his career. So the four cannons, which grateful Americans gave him as a token of appreciation for his help at Yorktown, ended up melted down for coin. Another guy at Yorktown also fought without pay: Henri de Saint Simon, at the time a noble nobody. After the American dust had settled, Saint Simon proposed a Panama Canal to the viceroy of Mexico ("no thanks"), returned to France, speculated wildly, lost all his money, turned socialist, went nuts, shot himself in the head (missed), and finally invented a new religion for business and industry (New Christianity), with plans for a world where science would take over from the church and society's rulers would be elected only from the ranks of entrepreneurs and bankers.

One New Christianity believer was, like Saint Simon, an aristocrat and canal freak by the name of Viscount Ferdinand de Lesseps, who was socially well placed and just getting over the disappointment of a promising political career untimely derailed by the usual government change of policy. However, since de Lesseps's dad had been responsible for choosing who would be ruler of Egypt back in 1800 (when Napoleon invaded the country and made that kind of magisterial decision), a generation later Ferdinand was good pals with Khedive Said, the guy now in charge—to whom Ferdinand suggested the Suez Canal. The Khedive couldn't get enough of it. A French-operated international waterway would raise Egypt's image from that of fleapit in one corner of the Turkish Empire to something a little more classy. In 1869, after ten years of digging by twenty-five thousand fellah laborers (and a couple of new steam-powered dredges), the canal opened with the biggest party since the pharaohs. Potentates and politicians arrived from all over Europe to enjoy the free champagne, together with freeloading lesser mortals like Dumas, Gauthier, Zola, Ibsen, and Thomas Cook—and de Lesseps's cousin, the beautiful Empress Eugénie of France (who'd talked her husband Napoleon III into backing the project in the first place).

Eugnie made her mark on history by setting the fashion for crinoline, wide-brim hats, cashmere, black pearls, and Louis Vuitton; wore a dress dyed with the new artificial aniline; was daughter of the Countess de Montijo who is said to have provided Bizet with the plot for *Carmen;* got more involved in French politics than she should have; and ended up with the mob baying for her blood as she was

somewhat of a cloud—that Paine whip up something to whip things up. Paine's *Common Sense* fit the bill. Living up to his name, Paine blew the whistle on some irregularities in the financial transactions between American government representatives and the French who'd spent so much to make sure independence happened. Back in Europe to recharge his revolutionary batteries (the shocks in France were yet to come), in 1791 Paine wrote the thing he's most famous for—*The Rights of Man*—and in 1792 the lesser known but much more important *The Rights of Man Part the Second,* in which he held forth on extraordinary ideas like the welfare state, progressive taxation of property, grant-aided education for the poor, and a state pension. Although many in England wanted to string him up for desertion, treason, defamation and being right, Paine's thoughts were music to the ears of Britain's radicals, one of whom helped Paine get *The Rights of Man* published in the face of opposition from high places.

This supportive type was William Godwin, as famous then as he is unknown now. Even contemporary radicals thought Godwin was a radical. He'd started out (as so many did) as a preacher, then turned to writing. In 1793 he was catapulted to fame by his *Political Justice,* a piece of crazy anarchist extremism that should have landed him in jail but for its calm, reasoned tone. Godwin rapidly got to know other left-wingers, including Wordsworth, Coleridge, and Lamb. He then married one of the first feminists, Mary Wollstonecraft, after getting her pregnant. Alas, a few months later, in 1797 the childbirth killed her. Godwin turned to novel writing and to looking for another wife. Neither did him much good. The novels failed and Mrs. Clairmont, the widow in question, seems to have been the original shrew. Godwin was perennially short of cash, so when he discovered that his-and-Mary's-daughter Mary's seducer (the poetic Shelley, with whom she eloped in 1813) was well heeled, he stopped fussing and started borrowing. In the end, the newlywed Shelleys had an idyllic but brief life together (he drowned) while traveling on the continent with Godwin's stepdaughter by the shrew, the nubile sixteen-year-old Claire Clairmont, who spoke fluent French (useful during the Shelley elopement). Claire threw herself at a famous poetic dissolute, introduced him to the Shelleys, had his child, chased him round Europe, and eventually gave up and went back to London with her pal, the (by-now-

escaping to England in 1870. Her husband (Napoleon Bonaparte's nephew) had made it to the throne by sheer "If at first. . . ." There were two coups: the one in 1836 failed, and he was exiled to the United States; the second, in 1840, failed, and he was exiled to the United Kingdom. Then he was elected president of France in 1848. Then in 1852 a referendum let him change the constitution, whereupon he declared himself emperor. Then for the next two decades, he got himself variously embroiled: fighting the Austrians on behalf of the Italians, fighting the Italians on behalf of the pope, joining in the Crimean War, trying to set up an extension of his empire in Mexico, and (biggest mistake) taking on the armed might of Prussia (lost, abdicated, and left for Chiselhurst, England, with Eugénie in tow). Late in his reign, Napoleon III had made tentative moves in the direction of democracy. Too little, too late. This also describes how he dealt with an appeal for help from a Hungarian nationalist (the guy wanted to take the "Austro" out of "Austro-Hungarian") named Kossuth, who had taken his cue from the rash of revolutions all over Europe in 1848 to demand parliamentary independence from Austro-Hungarian Imperial HQ in Vienna. He even got as far as becoming Hungarian Minister of Finance and issuing Hungarian money. His technique was uncomplicated: ignore Austria. It didn't work. In the end, let down by Napoleon III, he pushed off (as so many do) to exile and writing his memoirs.

Not long after his death in 1894, Kossuth's tactics became known as the "Hungarian policy" when an Irish political journalist wrote about him and suggested his Irish fellow countrymen take the same approach to their own political problems (in this case, for "Austro" read "Anglo"). To which end, journalist Arthur Griffith set up an organization called "We Alone" (Irish: *Sinn Fein)* and then an illegal parliament which uncomplicatedly ignored the Brits. In the very long run, after many vicissitudes including gunrunning, uprising, civil war, executions, periods of imprisonment, exile, and fund-raising in the United States (they collected six million dollars), the tactics paid off, and the British Dominion of the Free State of Ireland eventually became the Republic of Eire, with one of Griffith's early colleagues, the Spanish-Irish, American-born Eamonn de Valera, as its leader.

De Valera had been a math freak since the age when he could add and was at one point a math professor. Now, as head of government, he decided to set up the Irish equivalent of the Princeton Institute of Advanced Studies (where Einstein and other hotshot refugees had gone) and to invite his own hotshot science refugee. After secret meetings in Switzerland in 1939, anti-Hitler Nobel physicist Erwin Schrodinger (on the run from the Nazis) finally arrived in Dublin and stayed for seventeen years. In 1943 he decided to give a series of

widowed) Mary. They then made the equivalent of a dot-com bubble investment in an opera box (to rent out for profit) in a theater where the leading draw was diva soprano Jenny Lind, who was then persuaded by a bishop to withdraw from the stage, leaving Claire and Mary seriously out of pocket. At no point did Claire's poetic dissolute lover attempt to help.

That was not the kind of thing you did if you were Lord Byron, who had other mistresses, boyfriends, and an extravagant lifestyle to support—including a child he never met (he died first): the eccentric and mathematically talented Ada, born of his marriage to Isabella Milbanke. Ada was a sickly child who at age eleven produced a half-baked design for a flying machine. Then Mary Somerville, science popularizer and the greatest networker of the era, encouraged Ada's interest in math and turned her into the thinking man's fantasy. In 1833, when she was seventeen, Ada met Charles Babbage, inventor of the first computer, and they began a correspondence about everything. For nearly two decades, Ada and her compliant aristocrat husband, Earl Lovelace, did all they could to help Babbage get his project off the ground (they failed). In an 1843 article, Ada wrote presciently that such a machine might one day produce graphics, music, and be used for all kinds of practical purposes besides math. Part of Ada's passion with numbers also led her to gambling and serious debt. However, thanks to her title, looks, and brains, she mixed with the scientific elite, including Charles Wheatstone, professor of experimental science at London University, inventor of the concertina (his family made and sold musical instruments) and keen cryptographer. Wheatstone also measured the speed of current though a wire, exploded gunpowder with an electric signal, and generally stayed ahead of Samuel Morse. In 1837 Wheatstone and W. F. Cooke developed a telegraph (widely used by railroads in Britain, France, and Belgium) that activated two needles, which pointed at the letters being communicated. And a magnetoelectric dial telegraph. And a typewriting telegraph. Guess what Wheatstone cared about.

Wheatstone's telegraphic system was the one recommended for use between Australian state capitals by Charles Todd, Aussie superintendent of all things telegraphic. Todd had started out as an astronomical-tables calculator at Greenwich Observatory, London, graduating to

TRACK ONE

three general-public lectures about physics and biology. In the book that came out of the lectures (modestly titled *What is Life?),* Schrodinger wrote the magic words: "In calling the chromosome fibre a code-script we mean that its structure determines whether the egg would develop, under suitable conditions, into a black cock or into a speckled hen, into a fly or a maize plant, a rhododendron, a beetle, a mouse or a woman." In other words, was life encrypted?

END TRACK ONE

keeper of the electricity supply required for transmitting Greenwich Mean Time by telegraph. It was logical, therefore, that in 1855 he be chosen (young, no other prospects) to go out to south Australia and become their man in telegraphy (they had none at the time). By 1870 he'd laid so many miles of line that he was chosen to stretch one across the continent from Darwin (the arrival end of the cable from Britain) in the north to Adelaide in the south. This took just two years, thirty-six thousand poles, thirty-six thousand insulators, six lives, and more adventures than you could shake a boomerang at. On the way, they also set up a repeater station by a water hole in the middle of nowhere and named it, after Todd's wife, Alice Springs. The whole adventure made Todd famous and a Sir, then postmaster general, then government astronomer of Australia.

In 1889 his daughter Gwen married a young Anglo immigrant, William Bragg, professor of math and physics at Adelaide University. During his first eighteen years in the job, Bragg published just three minor papers. Then, suddenly, in 1904, he got interested in X-rays, and his life changed. Together with his son Lawrence, Bragg began to bounce X-rays off crystals, to analyze the patterns the rays then created when they interfered with each other after the bounce. This was supposed to say something about whether or not X-rays behaved like light waves. Instead, the Braggs suddenly saw everything the other way around. The patterns were actually revealing detailed information about the arrangement of the lattice of crystal atoms off which they had just bounced, which revealed the atomic structure and what the crystal was made of.

By 1951 British chemist Rosalind Franklin was using the technique (X-ray diffraction) to look at strands of DNA. What she saw were interference patterns that looked to her like "spinning four-bladed propellers." Closer inspection from all angles, plus math, gave her the idea that the DNA was structured in helical strands linked in some way.

END TRACK TWO

AND FINALLY . . .

Francis Crick was inspired by Schrodinger's book to search for the life code, and in 1953 he was shown (without her knowledge) Rosalind Franklin's collection of DNA X-ray diffraction pictures. All that remained was to find how a helical structure could contain the elements for a copying process that would allow the gene to replicate. Crick (and James Watson) came to the conclusion that if each of the four amino acids concerned (adenine, guanine, thymine, and cytosine) paired only with one other, then the helix could split and each pair would seek its partner and create another helix. This woud be the "code" read by some cellular mechanism that would translate it into orders for making the specific protein a cell needed for its structure and function. In 1962 Crick and Watson got a Nobel. Franklin, four years dead of cancer, did not.

TWENTY-THREE

1770: CHURCH SERMON TO HELICOPTER

Sitting where the Merrimack River flows into the Atlantic, the attractive town of Newburyport, Massachusetts, has a number of things to recommend it. Birthplace of the U.S. Coast Guard, thriving eighteenth-century commercial seaport, and site of shipyards that built many of the War of Independence American privateers (a.k.a. pirate ships), which made life difficult for the British navy. And (it is claimed) the site of a tea party before the Tea Party. Today, as one of the largest federally funded restoration areas in the United States, it probably looks a lot neater than it did back in 1770, which was when the population included two people who, in their different ways, changed the future.

TRACK ONE

At 6 A.M. on September 30, 1770, at the home of the Newburyport Presbyterian minister, a fifty-five-year-old visiting English preacher named George Whitefield had an attack of angina and died. In the fevered world of religious ranting, Whitefield was a media megastar before media megastars, though in private life a man of moderate tastes (his luxury of choice was cow heel). The century's greatest actor, David Garrick, said Whitefield could melt an audience with the way he said: "Mesopotamia." His voice was reputed to be audible a mile away. And despite the fact that he was stout and had a squint, this did not deter millions of people, in several countries, from standing in all weathers to hear him deliver over eighteen thousand sermons in thirty-four years. Some of the events were so inspiring that people fainted from salvation. At other times stones, rotten eggs, and dead cats were thrown. Love or hate Whitefield, nobody could ignore him. To such an extent that, over the years after his death, most of his bones were stolen, one at a time. All this, the result of the Great Awakening he had helped to cause.

Whitefield's own Great Awakening had happened in Oxford, back in 1732, when he met a fellow nerd who'd come up with a system for hitting the books that was so methodical his little stuffy study group became known as Methodists. Ironically, Charles Wesley caused Whitefield's Awakening six years before he, Wesley, had had his own. After which Wesley became the most prolific hymn writer of all time (no fewer than sixty-five hundred catchy numbers including "Hark the Herald Angels Sing"). This was after he'd lasted only a few months in the Georgia colony. His brother John (whose own Great Awakening was on the Wednesday following Charles's) did better, hanging on in Georgia over a year till a questionable relationship with a young woman drove him back to England. John became the real mover and shaker of Methodism, with more sermons (40,000) and miles (250,000) to his credit than even George Whitefield. By the time John died, there were more than 135,000 members, 541 preachers, and over 200 chapels saving the soul. John never made the chart toppers as a hymn writer, but his *Primitive Physic* (a do-it-yourself guide to disease, with treatments ranging from elderberry to electricity) ran to thirty-six editions. In the end he got a monument in Westminster Abbey, after (in 1789) a portrait by no less than George Romney.

By this time, Romney was well on his way from being a nobody from rural nowhere (via two years of study in Rome with a letter of introduction to the pope) to becoming the darling of the chattering classes. At the height of his fame (when he painted Wesley), Romney was knocking out no fewer than seven sitters a day. He did

TRACK TWO

Newburyport still boasts the house of a man who at the age of sixteen might well have been in George Whitefield's last congregation, goldsmith Jacob Perkins, who by 1798 had solved the major problem of the age: the new paper money was user-friendly but it was also forger-friendly. So Jacob clamped together up to sixty-four separate steel dies, each one of which printed a part of the banknote design. With this modular approach, clients could customize a single detail and in this way make their notes even harder to fake. By 1818 Jacob was in England, failing to sell the idea to the Bank of England (a) because it was full of bumbling idiots and (b) because a Brit inventor, Sir William Congreve, was taking up the bank's time with his idea for a fraud-proof colored watermark (produced by inserting dyed materials during the papermaking process). The counterfeiters couldn't replicate this process. Alas, as it turned out, nor could the bank.

Sir William had originally shot to fame with his Congreve rockets, most recently used to no effect in 1812 at an event immortalized by Francis Scott Key's reference to "rockets' red glare." Congreve rockets did little more than create confusion, as had been clear ever since their first trial by British navy captain Thomas Cochrane in 1809, during a battle against the French fleet. Nonetheless Cochrane did so well during the fight he made his admiral look bad. Then he went home and accused the Royal Navy of rampant corruption. Rocking the boat in these ways ensured Cochrane wasn't going to make it to the top in the near naval future, so in 1817, when he got an offer to take over the Chilean fleet, he jumped ship. By 1820 he was also running the Peruvian navy and making sure both countries ended up independent instead of far-flung bits of Spain. This endeared him to the regent of Brazil, Dom Pedro (full name too good to miss: Pedro de Alcântara Francisco Antônio João Carlos Xavier de Paula Miguel Rafael Joaquim José Gonzaga Pascoal Cipriano Serafim de Bragança e Bourbom), who was trying to achieve the same kind of breakaway from Portugal. Pedro gave Cochrane his navy. During exercises in the Atlantic, when some of the Brazilian ships revealed themselves to be little more than shipwrecks waiting to happen, Cochrane limped into a British port and wrote the regent a good-bye note.

everybody who was anybody: the archbishop of Canterbury, the Prince Regent's secret wife, duchesses galore. Part of his success may have been his prices: lower than most and less than they should have been. In his best two decades, he painted fifteen hundred people, some three times. One, however, he painted twenty-three times. An affair? Probably not. An infatuation? Certainly. The object of his brush and devotion was the mistress of a ne'er-do-well young aristocrat and about to become the mistress of the aristocrat's uncle, British minister to Naples Sir William Hamilton. When she later married the minister, she became Lady Emma. During her time in Naples, she also became confidante of the Queen of Naples, a spy for the British (she claimed), *poseuse extraordinaire* (in diaphanous outfits), indifferent singer, and most famously of all, lover of Admiral Horatio Nelson, the one-eyed, one-armed hero of Trafalgar—who fell for her hook, line, and sinker after he discovered she wore no underwear. Emma, William, and Horatio then became a ménage à trois scandal as they visited the courts of Europe on the way back to England for the cuckold's retirement.

This wasn't too bad for a semiliterate country girl who'd started life as a hooker and was, at one point, governess for Thomas Linley, the eighteenth-century version of Rodgers and Hammerstein. Like Emma, Linley was also from humble beginnings and had made his name as a singing teacher, harpsichordist, and concert director in the fashionable English country spa town of Bath. By 1774 his musical reputation and savings bought him a share in London's Drury Lane Theater and joint theater-manager's job with his son-in-law, who'd grown up with Linley's daughter, Elizabeth Ann (no mean accompanist herself), then eloped with and married her. Linley's music (for shows like *The Beggar's Opera, The Duenna,* and *The Tempest*) became popular in all the concert halls, especially after the success of the new son-in-law's greatest triumph, *The School for Scandal.*

Richard Brindsley Sheridan's *School* got rave reviews, was nearly banned, and broke box-office records. By 1777 Sheridan was so front-page he was even accepted into the ever-so-exclusive "club" on the say-so of that misanthrope's misanthrope, Dr. Samuel Johnson. Once inside, Sheridan rapidly became chummy with fellow-member politicos like Burke and Charles James Fox and in no time at all had graduated from scribbler to orator (once a shoo-in parliamentary seat had been arranged). In a world of florid overstatement, Sheridan outflowered them all, on one occasion (opening the six-year Warren Hastings trial for impeachment) holding forth, nonstop, for four days. As is so often the case in politics, Sheridan could briefly do no wrong (that was to come later), so he was one of the stars chosen in 1794 to meet and greet an important diplomatic visitor from the Austro-

TRACK TWO

In 1829 Pedro put his own imperial pen to paper in the interests of natural history. The object of Pedro's letter was to warn the Paraguayans of dire consequences if they didn't release from detention a vegetable-store owner, one Aimé Bonpland—who was, however, no ordinary retailer. His present troubles had arisen when he was in Argentina, just across the border from Paraguay, attempting to develop an international market for a local variety of green tea: yerba maté. Since Paraguay's dictator had the same kind of profit in mind, one of his snatch squads arrived to destroy Bonpland's yerba plantation, drag him over the border, stick him in the middle of the Paraguayan bush, and nip his little project in the bud. Why Pedro Emperor of Brazil should have bothered with the fate of a French expatriate botanist is because Bonpland had once been the gardener of Napoleon's Josephine, had written a major work on South American plants, was still a corresponding member of the French Académie des Sciences, and (last but extremely not least) from 1799 to 1805 had been one half of the greatest two-man exploration team the world had ever seen. The other half was Alexander von Humboldt, and together they had spent five years boldly going in South America. A few tidbits: while walking, riding, canoeing, and climbing more than six thousand miles, Humboldt and Bonpland visited Mexico, Cuba, Peru, Venezuela, Ecuador, and Colombia; went up the Orinoco; crossed the Cordilleras; peered down into volcanoes; collected sixty thousand plants; discovered the magnetic equator and the Humboldt Current; established hundreds of latitudes, longitudes, and heights so that more accurate maps could be made; invented climatology, ecology, and physical geography; and sent home enough bat droppings to start the German fertilizer industry. Later on, Humboldt did the same kind of thing in Siberia. And then, between being Prussian royal chamberlain and writing a thirty-four-volume record of his travels and a five-volume review of the Earth (modestly titled *Cosmos,* sales of which rivaled the Bible), Humboldt also wrote thousands of letters to the important and powerful.

One of these, English math whiz Charles Babbage, had written to Humboldt about not finishing something he was working on. This was true of almost everything Babbage did, and he did a lot: math, cryptography, life assurance, operational research, statistics, and inventions (too many to list but including shoes for walking on water and a plan for

Hungarian Empire, Prince Clemens von Metternich, soon to become the most powerful statesman in Europe.

For some strange reason, Metternich tends to bring glaze to the eyes of modern-day students at the mere mention of his name. For this some blame must be attached to his own description of his modus operandi: he bored people into doing what he wanted. But this devious, brilliant, manipulative latter-day Machiavelli must have had something else going for him, to judge by the number of beautiful, talented, and powerful women who fell. As regarded his day job, perhaps his greatest achievement was to breathe decades of extra life into the moribund monarchies of Europe (above all, that of his own boss the emperor Francis) with his invention of modern summitry: a series of international congresses (starting in 1814 with the Congress of Vienna) at which sovereigns, or the powers behind their thrones, could get together and exchange notes on how to stay sovereign with the aid of their own local versions of Metternich's new and efficient secret police.

The Prince died one year before his granddaughter became the first and most famous patron of a young German composer whose attachment to the old Germanic ways would have warmed Metternich's heart. But there was more to Richard Wagner than nationalism. His passions ranged from women to anarchism to nihilism to anti-Semitism to Germanic folk myth to Ludwig II of Bavaria (who paid all his debts and with whom he had a relationship that was so ambiguous even the public noticed). Wagner's music left its mark on composition for decades. And his operatic evocations of the mystic, ancient Teutonic legends triggered a response in the German psyche that was to last up to Hitler. Hitler was so taken by one of the operas (all about the search for the Holy Grail; subtext: the struggle to restore the purity of the Aryan race) that the führer exclaimed "Out of *Parsifal* I have made a religion."

Wagner's opera was to have a less well-known outcome. In 1923 *Parsifal* was staged at the Opera House in Madrid. Among the audience was a young engineer, Juan de la Cierva, whose spirits needed lifting because the project he'd been working on for the previous three years just wasn't getting off the ground. He found the solution to both problems on the stage that night. The set designer had included windmills, and it was the sight of their hinged sails that gave Juan the solution. He was trying to build a plane, which, if it went into a stall, wouldn't crash but would float down to the ground. To achieve this he had designed three narrow "wings," set on a central shaft above the fuselage, and spun round by the movement of the plane through the air. The trouble was that when a wing moved forward it generated lift and when it moved back it didn't, so the plane

a London-to-Liverpool speaking tube). His major obsession was triggered by seeing postrevolution French coiffeurs (out of work thanks to the new simpler hairdos) seconded to add and subtract the first trigonometry and logarithm tables for the new metric system—and making mistakes. So Babbage designed a dividing engine (a geared gizmo for addition and subtraction) and then an analytical engine (punched cards, stored program, fifty-decimal-place calculation). Neither machine ever went into operation, and he ran out of government grant money to complete this computer-before-the-computer in spite of inspired fund-raising attempts by Byron's daughter. Babbage also visited factories and wrote an incisive piece on manufacturing.

Babbage's piece influenced the century's greatest thinker on these matters, John Stuart Mill. One of those polymaths you love to hate. At age three, Greek; at seven, math; at thirteen, Latin, geometry, algebra, calculus, economics, logic, and lesser disciplines. By fourteen, he'd given himself a university education, with consequent damage to his personality. A party goer Mill was not. He always dressed in black. Mill's *Principles of Political Economy, System of Logic,* and *On Liberty* became the must-reads for any Victorian with pretensions to radicalism and understanding the social and philosophical issues of the day. In 1872 Mill became godfather to (and major influence on) the child that was to become world-class philosopher Bertrand Russell. About Russell, suffice it to say that he thought math was logic, invented logical atomism, and was clever enough to examine the incomprehensible Ludwig Wittgenstein for the latter's doctorate. Russell also wrote many books on philosophy for the "ordinary" reader, had four wives (the last one when he was eighty), numerous simultaneous mistresses, and founded the Campaign for Nuclear Disarmament.

Early on he taught and became friendly with John Maynard Keynes. Keynesian economics has pretty much run things since the mid–twentieth century (in a nutshell: raise taxes and lower spending during the good times; lower taxes and raise spending during the bad times). In 1918 Keynes fell for a Russian dancer, Lydia Lopukhova. She had already made her name in the Imperial Ballet in St. Petersburg and then spent several years in the United States, touring in vaudeville shows with her brother and sister as a specialty act (and putting ballet on the U.S. map). By the time Keynes met her, she was a famous prima ballerina,

tended to tilt in a manner that decidedly unimpressed pilots. Using the windmill idea, Juan put his wings on hinges so that the forward-moving wing could rise in the airflow and its airspeed would fall, while the backward-moving wing would do the opposite. This clever trick equalized the lift on both sides. So pilots with their engine out could land better than a brick. Good as it was, however, Juan's little idea never really made it big—but it made possible something that did.

END TRACK ONE

dancing major roles and taking London by storm. She stopped dancing soon after they married in 1925, and she settled down to look after the genius and spend time as an uncomfortable part of the self-regarding gaggle of Keynes's pals known as the Bloomsbury Group, who never quite came to terms with Keynes's marriage (complicated by the fact that he was gay and had a lover in Cambridge).

Lopukhova had made it to international stardom the way so many dancers did, working for one of history's great impresarios, Sergey Diaghilev. From the moment he hit the West in 1907 until he gave up dance for book collecting in 1927, Diaghilev spent every waking minute dreaming up ways to get funding for his independent ballet company, the only one around. Diaghilev changed ballet the way Copernicus changed astronomy (audiences at his productions felt the Earth move). This was ballet as nobody had ever seen it before, an extraordinarily individualistic, fantastic melding of dancer, music, and set. Diaghilev commissioned music from Stravinsky and sets from Picasso, Matisse, and Braque—and dazzled the West with the raw energy of Russian dance and composition. At one point, Diaghilev produced a Russian season in Paris, with the aim of introducing the new Russian nationalist school.

One of the talents on display was the pianist, conductor, and composer Rachmaninoff. When he lost everything during the Bolshevik Revolution, Rachmaninoff settled in the United States and became known worldwide for his *Rhapsody on a Theme of Paganini.* At a party in 1921, he met another Russian immigrant who was working in a Long Island barn trying to build airplanes from bits and pieces scavenged from junkyards. Rachmaninoff was so impressed by what he heard that he gave the man five thousand dollars and became vice president of his company. With this boost, Igor Sikorsky was able to build the great "Clipper" flying boats that flew the first regular Pan Am transatlantic passenger service.

END TRACK TWO

AND FINALLY . . .

In 1939, using Cierva's idea of variable-pitch rotors and his own expertise in airframe construction, Sikorsky built the first successful helicopter.

TWENTY-FOUR

1771: POTTERY TO NEON SIGNS

The terms of the first-ever money-back guarantee (in 1771 England) included the declaration that customers were "at liberty to return the whole or part of the goods . . . if they do not find them agreeable." You can hardly do better today. The shrewd business brains who thought up this unheard-of generosity were Josiah Wedgwood and Thomas Bentley.

TRACK ONE

Josiah Wedgwood (one-leg amputee, stammerer) dragged the pottery trade from the eighteenth to the twentieth century in one generation after he went into business in 1758. In 1765 his creamware china went over so big with Queen Caroline of England, Wedgwood asked for (and got) permission to call it "Queensware." Every aristocrat and wanna-be had to have it. And so did Catherine the Great of Russia, who ordered a service of 952 pieces, which the crafty Wedgwood exhibited (tickets only) before shipping. He also sent a thousand unsolicited buy-or-return samples to the German aristocracy and only got three returns. Not surprisingly he became well off and could indulge in his science passion by riding on full-moon nights to meet fellow members of the Lunar Society and do high-tech chat. The Society founder was reformed drunk Erasmus Darwin, grandfather of you-know-who, writer of bad poetry, and rumored to have vivified a plate of vermicelli by electric shock. Erasmus turned down an offer to become the king's doctor (hated London), knew the science panjandrums (Boulton, Watt, Priestley, et al.), and spent years writing up his theory that all organisms were descended from a single ancestor and that each generation passed on to the next generation what it had learned from its survival, which kind of set the scene for you-know-who.

In 1766, while on a botanizing expedition in Derbyshire, Erasmus bumped into another botany freak, who was renting a nearby house. Rousseau, the French philosopher, was already front-page news for his radical views on education (concentrate on the child) and social matters (the more complex a civilization gets, the more it corrupts people's innate qualities). Proof of the latter, he said, were Native Americans who ran a functioning and mature community without ever having heard of Greece or Rome or Descartes. "Noble savages," Rousseau called them. This down-with-artifice stuff got the French authorities annoyed enough for Rousseau to have to get out of town, and it also helped start the Romantic movement. Not surprising, then, that Rousseau gave good reviews to a new kid on the opera block, the Czech composer Christoph Gluck. Gluck in 1777 was getting ready for the Paris first night of his *Iphigenie,* one of several compositions that dumped the old conventions (which required that no matter what was going on in the plot, everything stopped for a ritornello or some other stylistically approved interlude). Gluck, instead, made the music fit the storyline and the character singing it, which shocked the traditionalists as much as if his performers had taken their clothes off on stage (which, in expressing these new naturalistic emotions, they metaphorically did).

Thomas Bentley spoke French and Italian, had traveled, and knew such big names as Sir William Hamilton, "collector" of newly excavated ancient Roman antiques. It was probably Hamilton's illustrated catalogs that Bentley showed to Wedgwood and got him going on neoclassical as opposed to baroque. The schmoozing Bentley also served as the front man in the Wedgwood London showrooms. Bentley had a business relationship with a man who made plaster casts and had a talented artist son. In 1775 the young John Flaxman joined Bentley and Wedgwood, sculpting cameos. After seven years of this (unsatisfying, badly paid), Flaxman lit out for Italy and was soon doing things for the expatriate community. The thing that made him an international star was a commission for seventy-three illustrations of the *Odyssey* and the *Iliad*. Flaxman went for the primitive feel of ancient Greek and Roman, and his simple outline figures became popular overnight. They were also easily copied, and though imitation was the sincerest form of flattery, it didn't pay, so he went home.

Back in the United Kingdom, the (now-religious) Flaxman found his form in the sentimental funeral monument, of which he did dozens in churches and cathedrals up and down the land (including Westminster Abbey and St. Paul's). The earlier commission to do the Greek-poetry illustrations, which had boosted his career, had come from an eccentric lady, Georgiana Hare-Naylor, cousin to the duchess of Devonshire. In England she always dressed in white, rode a white donkey, and was accompanied by a doe. In 1785, following elopement with an extremely dull writer of histories and plays (all of which failed), and after a couple of years in Rome (where she met and hired Flaxman), Hare-Naylor and her husband settled in the food center of the universe, Bologna, because they were short of cash and life was cheaper there. Hare-Naylor painted, became fluent in Greek (taught by an old Spanish Jesuit priest), and became pals with the priest's adopted daughter Clotilde Tambroni—writer of lesbian poetry and by this time professor of Greek at the university (famous for its female academics). In 1797 the Hare-Naylors headed back to England and an inheritance, leaving three boys in the care of Tambroni and the Jesuit. Tambroni eventually became famous as the greatest Greek scholar in southern Europe and, against all odds,

TRACK ONE

Round about this time, Gluck got a letter from a young man recently arrived in Paris after a brief naval career and who'd written an opera. Almost certainly you've never heard of this composer, Bernard Lacepède, since all his works were lost and anyway they all failed. Fortunately for him, Lacepède was good at science and the other letter he wrote (to Count Buffon, natural history boss of the period) got him an invitation to Buffon's museum and a commission for a definitive book on fish and whales. This effort got him the ear of Napoleon, and in 1804 Lacepede was on a committee of science mavens looking at what French scientific expeditions were bringing back from various distant spots. None was more distant than Tasmania, from which one shipload of twenty Gallic naturalists had collected one hundred thousand specimens including twenty-five hundred new species. One of the guys onboard was an artist named Lesueur, who, in 1815, bumped into a Scots-American geologist-explorer visiting Paris and went off as his naturalist secretary on a two-year geology tour of the Caribbean, the Allegheny Mountains, Florida, and Georgia. After which Lesueur did a spell as cartographer to the U.S. government, and then in 1825 rejoined his erstwhile traveling companion at the latter's utopian commune at New Harmony, Indiana, where Lesueur stayed and taught for ten years until he finally went back to France and died. By this time, his pal was also already long gone to Mexico. When they had originally met, the pal (William Maclure) was one of those people who'd "been there, done that," mixing successful business and equally successful rock tapping in Russia, Scandinavia, Scotland, France, Spain, Switzerland, and the Eastern United States. As a result of which, in 1817, he had produced the first geological map in America.

Earlier, in 1805 in Yverdon, Switzerland, he'd been impressed by the educational theories of Heinrich Pestalozzi, and he persuaded one of Heinrich's teachers (whom he met on the later Paris trip) to emigrate to the States. In 1825 the chap in question, Joseph Neef, was on the boat with Maclure and Lesueur, heading down the Ohio to New Harmony. Since his 1806 arrival in America, Neef had already opened schools in Kentucky and Pennsylvania, spreading the Pestalozzi gospel: teach children through real-life experience before books, give them plenty of outdoor exercise and music, and let them express themselves. This was radical stuff back then, and Neef was a little too forward-thinking for some of his pupils' parents. So a free-thinking socialist community like New Harmony suited him down to the ground and, except for a few years in Ohio, he taught there till he died. Meanwhile his Pestalozzian word had spread and caused trouble for one Henry Rogers, who joined the faculty of Dickinson College in 1829 and two years later was dismissed for his overliberal Pestaloz-

had a correspondence with Richard Porson, professor of Greek at Cambridge, who almost never replied to letters. Nor did he turn up for meetings. Or give any lectures. Or even live in Cambridge.

Porson's reputation rests on his editions of four Euripidean plays, since he never finished anything else. He was generally badly dressed, dirty, and drank a lot. And had some of the smallest handwriting ever seen. He picked other people's work to pieces and wrote countless paragraphs on bits of grammar and poetic meter—your average academic of the day. Even his marriage was a footnote, lasting all of five months before the lady died. She was the sister of James Perry, for whom Porson wrote a few pieces. Perry was editor of the *Morning Chronicle* and spent most of his day hanging around bars and coffeehouses picking up gossip and mingling with the literary crowd. At the time, the crowd included the poets and essayists Coleridge, Wordsworth, and Lamb and the economist Ricardo. All of them wrote for Perry at one time or another. Perry's dinner parties were famous, as were his lady-killer predilections. He was also so well liked that, although he was a friend of Sir William Hamilton, the fellow who later stood godfather to Perry's daughter was the very same cad who had cuckolded poor Hamilton: Admiral Lord Horatio Nelson, nation's hero for beating Napoleon.

One of Perry's young protégés was William Hazlitt, who started out as a popularizer of the "modern" philosophers (like Rousseau and Hume) and then graduated to essays and literary reviews for all the journals. The early-nineteenth-century London literary scene would never be the same again after this time: great poetry and prose from the likes of Byron, Shelley, and Walter Scott; great critical writing by Lamb, Hunt, and Hazlitt. Hazlitt, like many of them, was a literary hack, writing when he needed the money and drinking the rest of the time. His second attempt at marriage ended disastrously when, after returning ahead of his new wife from a continental trip, he wrote to ask when he should come back out to accompany her home and she wrote back to say don't bother. Hazlitt's run around Europe had taken in France, Italy, and Switzerland. In Florence he had bumped into Walter Savage Landor. Poet, Latinist, and perpetual schoolboy, Landor wrote the kind of stuff now only read by those in search of doctoral thesis material. To describe his poems as somewhere

zian sentiments. So Rogers took up with socialists and went to England for inspiration, only to find geology.

Back again in the United States, he did major surveys with his brother (who founded M.I.T.) and became hot stuff on hot stuff. His theory was that molten rock, upheaving zillions of years ago, had caused mountains to fold upon themselves to such a complicated extent as to end up with their layers, as it were, chronologically upside down. In 1844 Rogers was pleased to see his upheaval ideas emerging in the work of a Brit, William Hopkins, who explained the other mystery gripping everyone at the time. What were solitary giant boulders doing, lying on plains miles away from their origins? Mathematician Hopkins said: molten upheavals, causing massive floods (and here he applied the math) that could move a three-hundred-ton rock because the force of the current increased in ratio to the square of its velocity. He was wrong. It had been glaciers giving the shove. Undaunted, Hopkins pursued his molten ideas in experiments that eventually proved things get hotter, deeper. His 1851 tests involved squeezing molten materials, and in this he was aided by engineer William Fairbairn.

By now also an international shipbuilding and civil-engineering guru and adviser to governments, in 1859 Fairbairn was on a committee to find out why the 1858 transatlantic cable had failed after a few days and whether another one should be laid. With him in committee was Cromwell Varley, telegraphy expert, who was good on locating faults in submarine cables. In 1862 Varley built a simulator to reproduce submarine-cable conditions in which he could alter various parameters and simulate extra distance or failures. His tests showed that a properly protected cable ought to be able to transmit twelve words a minute across the Atlantic. This gave everybody the will to try again (successfully, in 1867). That year Varley was hired by William Orton, President of Western Union, to report on the condition of Western Union's telegraph lines. Varley was part of Orton's drive to make the company more proactive in supporting and adopting telegraphic innovations. Orton eventually contracted for Edison's multiplex telegraph, which would send four messages down the same wire. This was when Orton bumped into Gardiner Hubbard (a) because Hubbard was lobbying for a federal telegraph service to get prices down and (b) because Hubbard was funding a rival system that might do better than Edison's. Hubbard, a rich Bostonian, had a deaf daughter, Mabel, and while looking for somebody to teach her to speak in 1872, he met a young immigrant Scots speech teacher, Alexander Graham Bell, who soon fell in love with and married Mabel. And since Bell was also experimenting with telegraphs as an aid to the deaf, he got financial support from Hubbard for his own (potentially Edison-beating) multiplex system,

between stiff and sentimental would probably be generous. But his prose was delicate, and one work, *Imaginary Conversations,* was a great success. He had constant arguments with everybody, was well off and extravagant, and in 1835, after fourteen years in Florence, left his wife of the day in a huff, returning only twenty-three years later to live on the charity of his estranged family. It was there, during his final years, that he was smitten by a good-looking, fair-haired young American writer, Kate Field. Field had been sent to Italy to study music; wrote pieces for the *Boston Courier, Atlantic Monthly,* and *Scribner's;* and would end up correspondent for the New York, Chicago, and New Orleans newspapers, with a column of her own: "Kate Field's Washington." Field also met and charmed a famous English writer, Anthony Trollope, who was visiting relatives in Florence and with whom Kate formed a lifelong relationship.

By this time, Trollope had already hit the big time with novels like *The Warden* and *Barchester Towers,* and he knew all the top scribblers: Thackeray, George Eliot, Tennyson, Wilkie Collins. Trollope was also starting to make the kind of money that would let him give up his day job at the post office (but not before introducing the famous red mailboxes). A number of Trollope's novels were illustrated by John Millais, who back in 1846 at the age of sixteen had exhibited his *Pizarro Seizing the Inca of Peru* at the Royal Academy and then turned his back on the art establishment, setting up a group of artists (including Burne-Jones and Rossetti) dedicated to "what-you-see-is-what-you-get" realism based on the simple style of Italian frescoes painted before the time of Raphael. This bunch of esthetes called themselves the Pre-Raphaelite Brotherhood. The PRB was inspired by the writings of critic John Ruskin, who invited Millais to his home in Scotland for a holiday and whose wife, Effie Millais, Ruskin promptly stole. From the medieval, Millais progressed to paintings of children and then made a real fortune with portraits of prime ministers, royalty, and the aristocracy. In 1879 he went to see the work of another (amateur) painter, Sarah Bernhardt, a.k.a. the Divine Sarah, better known for her dramatic performances on stage and in bed. Millais's companion at the viewing was the Prince of Wales, soon to become Edward VII and one of Sarah's many lovers. Bernhardt, the illegitimate daughter of a high-class hooker who was also a prostitute to various French princes, was a smash hit from her first

TRACK ONE

which in 1875 made possible the telephone and the business Hubbard and Bell founded together.

Bell's telephone worked on extremely weak current. A few years later, the weakness turned out to be useful to a young French aristocrat, Arsene d'Arsonval, doctor turned physiologist. D'Arsonval inserted a frog muscle into a telephone circuit, and after a series of experiments, he announced that positive charges caused muscular elongation and negative charges contraction. He found that electricity generated heat in muscles—and invented physiotherapy. Early in his career, d'Arsonval developed instruments for measuring body heat, which led him to investigate low-temperature measurement. This was possible, he discovered, with a thermometer using superpurified gasoline, the only liquid that stayed liquid in the presence of liquid air. This was good news for gas liquefiers.

END TRACK ONE

day on stage with the Comédie Française in Paris and rapidly became a celebrity in Europe and America with her flamboyant acting, exotic costumes, and shocking lifestyle. Oscar Wilde wrote *Salome* for her, but the play was censored for blasphemy.

Back in 1864, when she was still unknown, Bernhardt (draped in velvet and nude behind a fan) had been snapped by the most famous photographer in France, Nadar (real name: Gaspard-Félix Tournachon). After working as secretary, shop assistant, smuggler, journalist, and spy, Nadar found fame with lithographed caricatures of the thousand most famous people in Paris. By the time he was photographing Bernhardt, his huge studio regularly housed photo parties where the subjects (Flaubert, Berlioz, Rossini, Verdi) sat, while their pals (Offenbach, Dumas, and others) fenced or drank and chatted. Nadar was obsessed by flight, and in 1863 he started the Society for the Encouragement of Aerial Locomotion by Lighter-than-Air Machines. The honorary president was Jules Verne, a man whom Nadar had met the year before at the Scientific Press Club and who admired Nadar so much that he put him in two novels, anagrammed as "Ardan."

Verne was encouraged by Nadar, who introduced him to his technical friends and gave the young novelist an introduction to the world of science. The following year, Verne wrote the highly successful *Journey to the Center of the Earth* (and then eighty more novels including *Twenty Thousand Leagues under the Sea* and *Around the World in Eighty Days).* Verne's remarkably accurate visions of the future (detailed in *The Diary of an American Journalist in 2890)* included high-rise skyscrapers, moving walkways, videophones, fax machines, and a global communications network. His ideas were to dominate science fiction from then on, in fast-paced adventure stories laced with science fact and an optimistic view of technology.

END TRACK TWO

AND FINALLY . . .

In 1902 Verne inspired George Claude, a pupil of d'Arsonval, to develop ways of distilling quantities of neon gas from the air, which Claude succeeded in liquefying. In 1910 Claude sealed some neon gas in a vacuum tube and sent in an electric charge. The tube glowed bright red. At the Paris Auto Convention that year, Claude's first neon lights went on display, to transform the world of advertising.

TWENTY-FIVE

1676: THEOLOGY TO SKYSCRAPER

In 1676 Samuel Butler's poem "The Elephant in the Moon" lampooned contemporary astronomy freaks (the astronomers see the lunar elephant, then discover there's a mouse in their instrument). The increasing availability of telescopes triggered arguments about the universe and, in particular, whether it was inhabited. This was theologically dangerous stuff, since the existence of aliens would mean Adam and Eve weren't everybody's progenitors (as the power-of-thumbscrew church said they were). Complicating the issue, Descartes had recently come up with a view of the cosmos filled with whirling vortices of matter, such as the solar system, raising the possibility that what happened here, could happen there. But with an eye to Rome, speculation regarding life elsewhere in the cosmos was muted (even Descartes said: "Dunno.").

TRACK ONE

In 1686 the first popular-science journalist, a Frenchman named Fontenelle (who also wrote general-reader stuff on politics, philosophy, mythology, medicine, and literature ancient and modern, and who produced the first-ever intelligible reports on Académie des Sciences activity) went out on a limb with *Plurality of Worlds.* The title ruled out any theological ducking and weaving, so Fontenelle got some very suspicious looks from Rome. But he got away with it on the basis of the argument that God could make anything possible (even something that broke God's own rules).

Two years later, Fontenelle's work went into English at the hand of the first female writer to earn her living by the pen. Afra Behn was quite an anomaly. Following a brief period in Suriname, she returned to England and wrote (over some years) the first story with a black hero: *Oroonoko.* After she married a Dutch merchant in London and was introduced to the royal court, her husband died and in 1666 Behn was sent on a spy mission to Holland. Her warning of an imminent Dutch attack up the Thames was ignored. A year later, Dutch warships arrived and destroyed the English fleet. Behn was then shipwrecked on the way home and, without money or husband, took up writing and became a succès d'estime with her novels and plays. In 1677 her *Rover* appeared on stage at the King's Playhouse. The players included a petite, shapely redhead, who had made it to the footlights via the casting couch. The starlet in question had begun her career as a hooker, then advanced to the post of orange seller, and finally, via a playwright's favors, got her first bit part in 1665—and from there, fame and fortune. Sir Christopher Wren, Samuel Pepys, and half of London fell for her. By the end of her stage career, her son was the duke of St. Albans and she was about to become Countess of Greenwich. Alas, Nell Gwynn never made it to the title, thanks to the untimely death of her lover, Charles II.

Charles II had also dispensed largesse to the illegitimate sons of his numerous other mistresses. His own career had had a bad start. After his father Charles I was beheaded by the Commonwealth authorities, he had spent several years in exile. In 1660 he returned to the throne and a country in need of fun and games after more than a decade of wet-blanket Puritans. Charles II reopened the theaters, handed out kingly rewards to all his pals, and stood godfather to the son of one of his faithful supporters. The child grew up to become Charles Townshend, high-powered politico. After sitting on the 1706 commission that pushed through the 1707 Union with Scotland, by 1721 Townshend was running British foreign affairs. He is probably best remembered by his nickname: Turnip. This, from his

TRACK TWO

In 1698 the great Dutch polymath Christiaan Huyghens joined in the is-there-life-out-there discussion with *Cosmothereos* (argument: other planets look the same; what other use would they have but to be inhabited; nature is varied enough to do it; yes, they're hotter or colder, but humans are adaptable; there seem to be clouds on Jupiter and atmosphere on Venus; a zillion stars could mean a zillion planets). And all this was from a Very Important Scientist: wave theory of light, rings of Saturn, major longitudinal studies, pendulum clock, telescope builder, and inventor (or not) of the clock spring. If he didn't, the inventor was Robert Hooke, another knew-everybody who built vacuum pumps for Boyle and who did early stuff with microscopes (compound eye of fly, structure of snowflakes), coined the term "cell," and invented barometers, a lens-grinding machine, windmill sail, air gun, odometer, and more. The clock-spring invention (Hooke said) happened in 1658 and the resultant spring-driven watch was built by his pal hi-tech quadrant maker Thomas Tompion. Tompion perspicaciously switched to watchmaking, manufactured over five thousand examples, became official watchmaker to the king, and developed a mechanism that enabled the watch to be flat instead of bulbous.

In 1711 Tompion took on a partner, George Graham, who then married Tompion's niece Elizabeth. Graham's great instrumental contribution was a tiny screw, angled against the edge of the turning baseplate of a telescope or theodolite, so you could turn the screw and move the instrument a fraction of an inch. In 1736 when the French decided to go to Lapland and Peru to measure one degree of latitude (Newton: "Earth is flattened at the Poles so a degree will be smaller up north than down south." Most French scientists: "Rubbish"), Maupertuis (pro-Newton and arrogant with it) took Graham's instruments and clock with him and proved Newton right. Maupertuis was another polymath—music, math, heredity, mining, optics—and, like everybody else, fascinated by the matter of where new organisms came from. Generally accepted at the time was that they were all there from the start (in which case Eve had two hundred thousand million embryos, inside each other, inside her). The alternate (Maupertuis's idea) was that organisms somehow "came together" when seminal

introduction around 1740 of the Norfolk Four-Course Rotation System, which consisted of planting the land in yearly succession with cereals, then turnips, then cereals, then grass. Townshend's novel idea of using turnips in year two avoided the usual practice of leaving the field fallow that year. An added bonus was that the turnips would feed livestock, whose manure would enrich the ground.

At the height of his fame, Townshend managed to arrange the release of an imprisoned scribbler who was leading a double life. As a journalist (sometimes libelous, hence the prison sentence), Daniel Defoe ran a three-days-a-week newspaper (he wrote all the articles himself) and was well known as a witty essayist. In 1719 he hit the big time with *Robinson Crusoe,* based on the story of a sailor four years marooned on Juan Fernandez Island. Then came his next best-seller, *Moll Flanders,* followed by several more popular titles. Meanwhile, behind closed doors, Defoe was a government spy recruited in 1706 to report on anti-English activists during the run-up to the 1707 Scotland-England Union. His spymaster was Robert Harley, Prime Minister from 1710, who also had the bright idea of privatizing the National Debt by giving owners of government bonds a 6 percent share in the South Seas Company, recently set up to trade in South America. After a slow start, company share prices went sky high. In January 1720, shares were at 128, and by August they were 1,000. By December the bubble had burst.

Just in time to bankrupt writer John Gay, who saw his precious thousand pounds' worth of South Seas stock rocket to 20,000 and then drop to 400. Gay knew Harley because they were both in the Scriblerus Club together with other literati like Swift and Pope. Gay finally made it big with his 1728 *Beggar's Opera,* which broke all box-office records. Produced by Lincoln's Inn Theater manager John Rich, the play, it was said, "made Gay rich and Rich gay." But such moments of income were few and far between. The only other piece that ever made Gay any money was a book of fables published in 1727 and written for Prince William, duke of Cumberland, later to become known as "the butcher" after the way he behaved in Scotland following the Battle of Culloden in 1746. The encounter had been the last fling of Prince Charles Stuart (a.k.a. Bonnie Prince Charlie), who claimed the English throne. Cumberland (eight thousand men and artillery) beat Charles (five thousand men and claymores) and then went on a rampage through the Scottish Highlands, looting, burning, and slaughtering wherever he went. Culloden was Cumberland's only victory. In all his other major battles, facing any opposition as well equipped and trained as his own, he lost.

At one of these failures (Fontenoy, 1745), a young officer named Robert Monckton served under Cumberland. Monckton went on to

fluid from male and female brought the necessary bits in contact. In 1759 this embryonic thought was developed by a German, Caspar Wolff, who suggested all the bits were already there and solidified out of a "fluid" present in all organisms. Based on his observation of cabbage, Wolff said you could see the little growth points where new leaves and flowers would come from. In 1790 this concept was picked up by Goethe in his spare time from being a major novelist, poet, playwright, philosopher, and genius. Goethe wasn't bothered by the nitty-gritty and the observation of detail and listing of parts and all that humdrum research stuff. Being Goethe, he sought a Grand Unified Theory of Life. His answer to "How did organisms know what to become?" was that in the mind of Nature there were what he called "Ur-organisms" (the drawing-board designs, so to speak), and all forms of an organism that came into subsequent existence were variations on this theme. This was why a plant leaf bud "morphed" into all the separate bits of a new plant.

In 1821 Goethe dashed off a quick translation of an "Ode on the Death of Napoleon" by Italian Alessandro Manzoni, whose reputation he'd been fostering. The Ode (also translated by Wordsworth, Byron, Lamartine, and others) caused a sensation partly because memories of Napoleon were still fresh and more so because Napoleon had been the archenemy of the Austrians who were now repressively occupying Italy (and, therefore, were archenemies to all Italians). Manzoni's anti-Austrian nationalism was powerfully expressed in his read-between-the-lines novel *I Promessi Sposi.* Like several other of his major works, it was a story set in the past in an "occupied country." Everybody got the point, but there was nothing the Austrian censor could do about "historical" material, was there? (Verdi would pull the same trick later on and get away with it, too.)

Manzoni might have taken some of his reformist views from his grandfather, Cesare Beccaria, the first real criminologist. Back in 1764 Beccaria had written his masterwork: *Crime and Punishment,* a detailed study of how capital punishment and torture didn't help reduce violence and stealing. So how about fitting the punishment to the crime? How about education and rehabilitation? How about more, better-staffed, and cleaner prisons? In 1767, when Voltaire wrote a blurb for the French version, Beccaria's book had became an international best-seller and Beccaria a celebrity. Governments all over Europe started taking up

more success than his boss had enjoyed. In 1752 he was sent to Nova Scotia as part of the plan to drive the French out of Canada and was second-in-command during the attack on Quebec in 1759, during which he was wounded and his commander General Wolfe killed. By 1761 Monckton was governor of New York. In 1762 he took the Caribbean island of Martinique from the French. In 1763 he returned to England, was promoted to general, led a quiet life, and achieved immortality in one of the most shocking paintings of the century (because it showed him and the dying Wolfe and all the other officers in the picture wearing army uniforms instead of Roman garb). *The Death of Wolfe* was first shown at the Royal Academy in London in 1771 and caused a sensation because it was the first modern-dress historical painting of a recent major event. The work made the artist, Benjamin West, an instant household name. West was an American who had arrived in London eight years earlier for a few months' visit but stayed for fifty-seven years, becoming president of the Royal Academy, history painter to the king, and a man who never put a foot wrong, with one exception: His 1783 proposal for a series of works on (recent) American Revolutionary subjects went over like a lead balloon.

In 1784 another young American painter fetched up in London, and West took him under his wing. The encouragement didn't have much effect. In 1791 Robert Fulton saw his first English canal, promptly switched career paths to civil engineer, and started churning out plans for inclined planes, aqueducts, bridges, and other matters associated with waterborne transportation. In Paris to promote his new work, he fell in with boat makers and proceeded to build the first submarine. It failed. His next stop was steamboats. With the support of the well-endowed Robert Livingston (U.S. minister to France), Fulton had a 3-mph paddle steamer on the Seine by 1802. With a better engine he went back to the States, and by 1807 the North River Steam Boat was running once a week between Albany and New York. Fulton's better engine had been built for him by James Watt, who had the original steam-power idea suggested to him back in 1758, when he'd met John Robison, professor at Glasgow University. Robison's pal Adam Smith then got Watt the job of university instrument maker, and while repairing an old Newcomen steam pump, Watt introduced some improvements and kicked off the Industrial Revolution.

Robison's son, a pal of the younger Watt, made a ton of money in British India and came home to a leisured life of inventing (screw-cutting improvements, cheaper photographic techniques, an article on metal-turning in the *Encyclopedia Britannica*) and involvement in the Edinburgh Royal Society. At some point, on the banks of the Forth and Clyde canal, he also carried out experiments on how water flow

his suggestions, except for the Brits. Thanks to the Industrial Revolution, Britain's cities were jammed with poverty-stricken immigrants from the countryside. At the same time, the new industrial rich were ostentatiously flaunting their wealth. The ensuing crime wave was met with the full force of the law. This included hanging for offenses like shoplifting, picking pockets, and "consorting for more than a month with persons calling themselves Egyptians."

Member of Parliament Sam Romilly, influenced by Beccaria's ideas, led the fight for penal reform and finally moved things along. One of his major successes lay in persuading Home Secretary Robert Peel that London needed some kind of security organization to help keep citizens' throats from being slit. In 1829 Peel set up the Metropolitan Police Force, nicknamed (after Peel's first name) "bobbies." Ironically, Romilly (a morose type) finally slit his own throat. Peel was Prime Minister twice. In 1834, during his first administration, he appointed Alexander Baring to settle a ticklish matter with the United States: a dispute over bits of the U.S.-Canada border. The argument had, at one point, involved an American "General" Van Rensslaer and some volunteer pals who crossed the border at Niagara on the good ship *Caroline* and got themselves fired at by the Brits. Alexander Baring was the recently retired head of Baring Bros. (the bank that had financed the Louisiana Purchase) and was married to the daughter of a U.S. senator. Moreover U.S. Secretary of State Daniel Webster had been legal adviser to the Baring Bank for three years. So Anglo-American negotiations were expected to go well, which they did. Part of the outcome was a title for Baring, turning him into Lord Ashburton—which is why the paperwork resulting from the discussions is now known as the Webster-Ashburton Treaty, thanks to which the United States got a bit of Canada.

Things were also helped along by the U.S. minister in London at the time. Edward Everett (like Webster, an Ivy League type) had started out with a Harvard degree in divinity, spoke seven languages, and spent years in Europe during which he got a doctorate from Gottingen. This was followed by the chair of Greek literature at Harvard, marriage to one of the better families of New England, and election to Congress. In 1835 he had become governor of Massachusetts. What next but the London job? In 1841 Webster sent Everett there as minister, and he charmed the pants off everybody with his

TRACK ONE

affected boats moving at different speeds. By 1832 this had qualified him to help design ships using water-resistance math and to join John Scott Russell's committee on waves. In 1849 another of Russell's committees awarded a Society of Arts gold medal for a hot new method for pressing sugar cane, developed by the inventor Henry Bessemer. Bessemer's tinkering then led him to something even hotter: the discovery that blowing air through molten iron would create steel.

END TRACK ONE

scholarship and aristocratic savvy. In 1844 he received an unexpected and importunate letter from a brash American who was looking for the right introductions and who had a midget in tow. The midget did so well that he (Tom Thumb) and the brash person in question (P. T. Barnum) got to have tea with Queen Victoria. Barnum was known as the Prince of Humbug. He'd started out with a grocery store, a newspaper, and a boardinghouse, then graduated to a freak show (bearded lady, flea circus, woolly horse, mermaid, etc.), from there to Congress, lecturing sporadically on temperance. Finally, in 1871 he opened "The Greatest Show on Earth," touring 140 towns a year from Nova Scotia to California.

In 1854 a young man auditioned an act for Barnum that looked like public suicide. Before meeting Barnum, the young man had moved around New York and Vermont working as a mechanic, sawmill owner, and carriage maker. Finally he turned to bedstead manufacture, and in 1852 he was in charge of relocating his company from New Jersey to Yonkers, New York. During the installation of machinery into the new factory, he developed a hoist for lifting various bits of equipment. The hoist also incorporated a ratchet so that if the hoist rope broke the machinery wouldn't fall. Early interest in developing the idea was sufficient for him to cancel plans for gold prospecting in California. His big break came thanks to Barnum, who was managing the American Institute Fair at the New York Crystal Palace Exhibition and offered him a spot in the show. The "suicide" act was to hoist himself forty feet up on a platform and then deliberately and dramatically (to gasps of horror) cut the hoist rope. To the astonishment of all, the platform remained in place. Elisha Otis had made his point, and as a result, from 1856 his career went onward and upward (and downward) as the manufacturer of the first elevator.

END TRACK TWO

AND FINALLY . . .

In 1884 the Otis elevator and Bessemer's steel were just what William Jenney needed to make possible the first skyscraper, the twelve-floor Home Insurance Company building in Chicago.

BIBLIOGRAPHY

Aitken, Hugh G. *Syntony and Spark: The Origins of Radio.* New York and London: John Wiley & Sons, 1976.

Ammon, Harry. *The Genet Mission.* New York: W. W. Norton & Co. Inc., 1973.

Baines, Paul. *The House of Forgery in Eighteenth-Century Britain.* Aldershot: Ashgate, 1999.

Baker, Keith Michael. *Condorcet: From Natural Philosophy to Social Mathematics.* Chicago and London: University of Chicago Press, 1975.

Bastin, John. "Memoir of Thomas Horsfield." In *Zoological Researches in Java, and the Neighbouring Islands,* by Thomas Horsfield. Singapore: Oxford University Press, 1990.

Batchelor, George. *The Life and Legacy of G.I. Taylor.* Cambridge: N.p., n.d.

Baugh, Christopher. *Garrick and Loutherbourg.* Cambridge: Chadwyck-Healey, 1990.

Beeson, David. *Maupertuis: An Intellectual Biography.* Oxford: Voltaire Foundation, 1992.

Beik, Paul H. *Louis Philippe and the July Monarchy.* Princeton, N.J.: D. Van Nostrand Company, Inc., 1965.

Bierbrier, M. L. *Who Was Who in Egyptology.* London: Egypt Exploration Society, 1995.

Bierman, John. *Napoleon III and His Carnival Empire.* London: John Murray, 1989.

Borgman, Albert S. *Thomas Shadwell: His Life and Comedies.* New York: New York University Press, 1928.

Brock, William H. *Justus von Liebig: The Chemical Gatekeeper.* Cambridge: Cambridge University Press, 1997.

Brown, Frederick. *Zola: A Life.* London: Papermac, 1997.

Burnett, Graham D. *Masters of All They Surveyed: Exploration, Geography and a British El Dorado.* Chicago and London: University of Chicago Press, 2000.

Cashin, Edward J. *Governor Henry Ellis and the Transformation of British North America.* Athens and London: University of Georgia Press, 1994.

Chandler, S. B. *Alessandro Manzoni: The Story of a Spiritual Quest.* Edinburgh: Edinburgh University Press, 1974.

Childs, Virginia. *Lady Hester Stanhope: Queen of the Desert.* London: Weidenfeld & Nicolson, 1990.

BIBLIOGRAPHY

Church, Leslie F. *Oglethorpe: A Study of Philanthropy in England and Georgia.* London: Epworth Press, 1932.

Cikovsky Jr., Nicolai. *George Inness.* New York and London: Praeger Publishers, 1971.

Cohen, Ernst. "Jacobus Henricus van't Hoff." In *Great Chemists.* Edited by Eduard Farber, New York: Interscience Publishers, 1961.

Costello, Peter. *Jules Verne: Inventor of Science Fiction.* London: Hodder & Stoughton, 1978.

Crickmore, Paul F., and J. Alison. *F-117 Nighthawk.* Osceola: MBI Publishing Company, 1999.

Crosland, Margaret. *Louise of Stolberg, Countess of Albany.* Edinburgh and London: Oliver & Boyd, 1962.

de Marly, Diana. *The History of Haute Couture, 1850–1950.* London: B. T. Batsford Ltd., 1980.

Demers, Patricia. *World of Hannah More.* N.p., 1996.

Dick, Steven J. *Plurality of Worlds: The Origins of the Extraterrestrial Life Debate from Democritus to Kant.* Cambridge: Cambridge University Press, 1982.

Dunbar, Janet. *Peg Woffington and Her World.* London: Heinemann, 1968.

Dwyer, T. Ryle. *Eamon de Valera.* N.p., 1980.

Edgerton, Harold E. & James R. Killian, Jr. *Moments of Vision: The Stroboscopic Revolution in Photography.* Cambridge, Mass.: MIT Press, 1979.

Eisen, Cliff. "Salzburg under Church Rule." In *The Classical Era: From the 1740s to the End of the 18th Century.* Edited by Neal Zaslaw. London: Macmillan, 1989.

Feinstein, Elaine. *Pushkin.* London: Weidenfeld & Nicolson, 1998.

Field, Kate. *Kate Field: Selected Letters.* Edited by Carolyn J. Moss. Carbondale and Edwardsville: Southern Illinois University Press, 1996.

Fitzlyon, April. *The Price of Genius: A Life of Pauline Viardot.* London: John Calder, 1964.

Fitzsimons, Raymund. *Barnum in London.* London: Geoffrey Bles, 1969.

Fothergill, Brian. *The Cardinal King.* London: Faber and Faber, 1958.

Gardner, Martin, ed. *The Annotated "Night Before Christmas."* New York: Summit Books, 1991.

Gervaso, Roberto. *Cagliostro: A Biography.* London: Victor Gollancz Ltd., 1974.

Gittings, Robert, and Jo Manton. *Claire Clairmont and the*

Shelleys, 1798–1879. Oxford and New York: Oxford University Press, 1992.

Giustino, David de. *Conquest of Mind: Phrenology and Victorian Social Thought.* London: Croom Helm, 1975.

Glendinning, Victoria. *Trollope.* London: Hutchinson, 1992.

Goebel, Julius. *The Struggle for the Falkland Islands: A Study in Legal and Diplomatic History.* New York and London: Kennikat Press, 1927.

Gold, Arthur, and Robert Fizdale. *The Divine Sarah: A Life of Sarah Bernhardt.* London: HarperCollins Publishers, 1992.

Gosling, Nigel. *Nadar.* London: Secker & Warburg, 1976.

Gutman, Robert W. *Mozart: A Cultural Biography.* London: Secker & Warburg, 2000.

Hardwick, Mollie. *Emma, Lady Hamilton.* London: Cassell, 1969.

Hodges, Sheila. *Lorenzo da Ponte: The Life and Times of Mozart's Librettist.* London: Granada, 1985.

Hopkins, Graham. *Nell Gwynne.* London: Robson Books, 2000.

Horowitz, Joseph. "Dvorak and the New World: A Concentrated Moment." In *Dvorak and His World.* Edited by Michael Beckerman. Princeton, N.J.: Princeton University Press, 1993.

Howarth, William D. *Beaumarchais and the Theatre.* London and NY: Routledge, 1995.

James, Patricia. *Population Malthus: His Life and Times.* London: Routledge & Kegan Paul, 1979.

James, T. G. H. *Howard Carter: The Path to Tutankhamun.* London and New York: Kegan Paul International, 1992.

Jenkins, Elizabeth. *The Shadow and the Light: A Defence of Daniel Dunglas Home, the Medium.* London: Hamish Hamilton, 1982.

Jensen, Ronald J. *The Alaska Purchase and Russian-American Relations.* Seattle and London: University of Washington Press, 1975.

Johnson, Douglas. *Guizot: Aspects of French History, 1787–1874.* London: Routledge & Kegan Paul, 1963.

Joll, James. *The Anarchists.* London: Eyre & Spottiswoode, 1964.

Jones, George Fenwick. *The Salzburger Saga: Religious Exiles and Other Germans along the Savannah.* Athens: University of Georgia Press, 1984.

Keates, Jonathan. *Purcell: A Biography.* London: Pimlico, 1995.

Kelly, Linda. *Juniper Hall: An English Refuge from the French Revolution.* London: Weidenfeld & Nicolson, 1991.

Keynes, Milo, ed. *Lydia Lopokova.* London: Weidenfeld and Nicolson, 1983.

Knif, Henrik. *Gentlemen and Spectators: Studies in Journals, Opera and the Social Scene in Late Stuart London.* Helsinki: Finnish Historical Society, 1995.

Leppmann, Wolfgang. *Winckelmann.* London: Victor Gollancz Ltd., 1971.

Lewis, Lesley. *Connoisseurs and Secret Agents in Eighteenth-Century Rome.* London: Chatto & Windus, 1961.

Longworth, Philip. *The Art of Victory: The Life and Achievements of Generalissimo Suvorov (1729–1800).* London: Constable, 1965.

Lynch, Kathleen M. *Jacob Tonson, Kit-Kat Publisher.* Knoxville: University of Tennessee Press, 1971.

Mabberley, D. J. *Jupiter Botanicus: Robert Brown of the British Museum.* Braunschweig: Verlag von J. Cramer; London: British Museum, 1985.

Mack, John E. *A Prince of Our Disorder: A Life of T. E. Lawrence.* Cambridge, Mass., and London: Harvard University Press, 1998.

Mackaness, G. *Fourteen Journeys over the Blue Mountains of New South Wales, 1813–1841.* Sydney: N.p., 1950.

Mackay, Thomas. *The Life of Sir John Fowler, Engineer.* London: John Murray, 1900.

Maestro, Marcello T. *Cesare Beccaria and the Origins of Penal Reform.* Philadelphia: Temple University Press, 1973.

Manton, J. *Mary Carpenter and the Children of the Streets.* London: Heinemann, 1976.

Marshall-Cornwall, James. *Marshal Massena.* London: Oxford University Press, 1965.

McCarthy, John A. *Christoph Martin Wieland.* Boston: Twayne, 1979.

Mehren, Joan von. *Minerva and the Muse: A Life of Margaret Fuller.* Amherst: University of Massachusetts Press, 1994.

Miller, F. Fenwick. *Harriet Martineau.* London: Kennikat Press, 1972.

Moore, Walter. *A Life of Erwin Schrodinger.* Cambridge: Cambridge University Press, 1994.

Morgan, Augustus de. *Newton: His Friend and His Niece.* London: Dawsons, 1969.

Murphy, Antoin E. *John Law: Economic Theorist and Policy-Maker.* Oxford: Clarendon Press, 1997.

Murray, Alexander, ed. *Sir William Jones, 1746–1794: A Commemoration.* Oxford: Oxford University Press, 1998.

Neeley, Kathryn A. *Mary Somerville: Science, Illumination, and the Female Mind.* Cambridge: Cambridge University Press, 2001.

Nokes, David. *John Gay: A Profession of Friendship.* Oxford: Oxford University Press, 1995.

Oldroyd, David R. "Geological Controversy in the Seventeenth Century: Hooke vs. Wallis and Its Aftermath." In *Robert Hooke: New Studies.* Edited by Simon Schama. Hampshire: Boydell Press, 1989.

Oman, Carola. *Nelson.* London: Greenhill Books, 1996.

Orton, Diana. *Made of Gold: A Biography of Angela Burdett Coutts.* London: Hamish Hamilton, 1980.

Palmer, Alan. *Metternich.* London: Weidenfeld & Nicolson, 1972.

Piggott, Patrick. *The Life and Music of John Field, 1782–1837, Creator of the Nocturne.* London: Faber and Faber, 1973.

Pole, William, ed. *Life of Sir William Fairbairn, Bart.* Newton Abbot: David & Charles Reprints, 1970.

Posner, Donald. *Antoine Watteau.* London: Weidenfeld & Nicolson, 1984.

Ralph, Robert. *William Macgillivray.* London: HMSO, 1993.

Read, Donald. *Feargus O'Connor: Irishman and Chartist.* London: Arnold, 1961.

Reich, Nancy B. *Clara Schumann: The Artist and the Woman.* London: Victor Gollancz Ltd., 1985.

Richardson, Joanna. *Baudelaire.* London: John Murray, 1994.

Robbins Landon, H. C. *Haydn: The Years of The Creation, 1796–1800.* London: Thames and Hudson, 1977.

Robertson, Alec. *Dvorak.* London: J. M. Dent & Sons Ltd., 1964.

Ross, Ian Simpson. *The Life of Adam Smith.* Oxford: Clarendon Press, 1995.

Ross, Michael. *The Reluctant King: Joseph Bonaparte, King of the Two Sicilies and Spain.* London: Sidgwick & Jackson, 1976.

Ross, M. J. *Polar Pioneers: John Ross and James Clark Ross.* Montreal and London: McGill-Queen's University Press, 1994.

Rowe, F. M. "The Life and Work of Sir William Henry Perkin." *The Journal of the Society of Dyers and Colourists* 54, no. 12 (December 1938): 551–562.

Royde-Smith, Naomi. *The Double Heart: A Study of Julie de Lespinasse.* London: Hamish Hamilton, 1931.

Russell, Colin A. *Edward Frankland: Chemistry, Controversy and Conspiracy in Victorian England.* Cambridge: Cambridge University Press, 1996.

Salaman, Redcliffe N. *The History and Social Influence of the Potato.* Rev. ed. Cambridge: Cambridge University Press: 1985.

Sayre, Anne. *Rosalind Franklin and DNA.* New York: Norton, 1975.

Shepherd, John. *The Crimean Doctors: A History of the British Medical Services in the Crimean War.* Vols. 1 and 2. Liverpool: Liverpool University Press, 1991.

Shionoya, Yuichi, ed. *The German Historical School: The Historical and Ethical Approach to Economics.* London and New York: Routledge, 2001.

Slaughter, M. D. *Immortal Magyar: Semmelweis, Conqueror of Childbed Fever.* New York: Henry Schuman, 1950.

Soltau, Roger H. *The Duke de Choiseul.* Oxford: B. H. Blackwell, 1908.

Sowell, Thomas. "Sismondi: A Neglected Pioneer." In *Henry Thornton (1760–1815), Jeremy Bentham (1748–1832), James Lauderdale (1759–1839), Simonde de Sismondi (1773–1842).* Edited by Mark Blaug. Aldersh: Edward Elgar Publishing Limited, 1991.

Stark, Tony. *Knife to the Heart: The Story of Transplant Surgery.* London: Macmillan, 1996.

Steinlen, A. *Charles Victor de Bonstetten.* Lausanne, N.p., 1860.

Stevenson, Robert Louis. *Memoir of Fleeming Jackson.* London: Heron Books, 1969.

Stewart J., Douglas. *Sir Godfrey Kneller and the English Baroque Portrait.* Oxford: Clarendon Press, 1983.

Strakosch, George R., ed. *The Vertical Transportation Handbook.* New York and Chichester: John Wiley & Sons, 1998.

Tarling, Nicholas. *The Burthen, the Risk and the Glory: A Biography of Sir James Brooke.* Oxford: Oxford University Press, 1982.

Thomson, Alice. *The Singing Line.* London: Chatto & Windus, 1999.

Thrasher, Peter Adam. *Pasquale Paoli: An Enlightened Hero, 1725–1807.* London: Constable, 1970.

Todd, Janet. *The Secret Life of Afra Behn.* London: Pandora, 2000.

BIBLIOGRAPHY

Travers, Morris W. *A Life of Sir William Ramsay.* London: Edward Arnold (Publishers) Ltd., 1956.

Troyat, Henri. *Turgenev.* London: W. H. Allen & Co., 1991.

Turner, Paul V. *Joseph Ramee: International Architect of the Revolutionary Era.* Cambridge: Ccambridge University Press, 1996.

West, Richard. *The Life and Strange, Surprising Adventures of Daniel Defoe.* London: Flamingo, 1998.

Whitworth, Rex. *William Augustus, Duke of Cumberland.* London: Leo Cooper, 1992.

Williams, John R. *The Life of Goethe: A Critical Biography.* London: Blackwell, 1998.

Witte, William. *Schiller.* Oxford: Basil Blackwell, 1949.

Wright, Helen. *Explorer of the Universe: A Biography of George Ellery Hale.* New York: American Institute of Physics, 1994.

Yeo, Richard. *Encyclopaedic Visions: Scientific Dictionaries and Enlightenment Culture.* Cambridge: Cambridge University Press, 2001.

Yuknis, Anthony D. *Thaddeus Kosciuszko: The Champion of Freedom.* N.p.: J.K. Tautmyla, 1966.

Zachar, G. Pascal. *Endless Frontier: Vannevar Bush, Engineer of the American Century.* Cambridge, Mass.: MIT Press, 1999.

INDEX

Abel, Fred, 143
Abel, Leopold August, 143
Academy of Science (French), 147, 229, 246
acting styles, 66, 166, 167, 208, 226, 243
Adam, Robert, 176–78
Adams, John Quincy, 149–51
Adams, Steven (Michael Maybrick), 41–43
Adams, Thomas, 203, 204
Addison, Joseph, 96, 116, 197, 199
aeronautics, 2, 53–54, 230–32, 234
agar-agar, 142, 144
Agassiz, Alexander, 210
Agnesi, Maria Gaetana, 126
agricultural chemistry, 63
air-cushion theories, 193–94
Airy, George, 47
Alamo (Mission San Antonio de Valero), 203
Alaska, U.S. purchase of, 19
Albani, Alessandro, 119, 176
Albert of Saxe-Coburg-Gotha, Prince, 202
Alembert, Jean le Rond d', 36, 89, 201
Alfieri, Vittorio, 121, 156
Algarotti, Francesco, 199
alginates, 160–62
Alps, 163
American Coast Survey, 108
American Museum of Natural History, 70
American War of Independence, 38, 69, 168, 187–89, 225
  French support of, 78–80, 81, 137, 159, 183, 206–8, 216–18, 219
Amherst, Jeffrey, 137–39, 141
ammonia, 120, 123
anchors, 53
Andersen, Hans Christian, 209
Andrews, Stephen Pearl, 188
anesthetic, 40, 120
Anglican church, 8, 18
Anglo-Saxon, 101
aniline dye, 123, 143, 218
Antarctic expedition, 28
anthracene, 123, 124
anthropology, 117
antiaircraft guns, 73–74
anticyclones, 111
anti-Semitism, 18, 67, 230
antisepsis, 42, 120, 153, 160
Antommarchi, Francesco, 150
aphids, 147
Arbuthnot, John, 116
archaeology, 9, 50, 58, 71, 101
architecture, 68, 69–71, 76, 91, 98, 171–73, 176–78, 200, 254
Ariosto, Ludovico, 56
Aristotle, 1
Arlington National Cemetery, 91
Army, U.S., professionalization of, 46
Arnim, Achim von, 67
Arnold, Joseph, 26
Arouet, François-Marie (Voltaire), 36, 79–81, 89, 130, 177, 199, 249
arsenic, 43
art, *see* painting
art collections, 76, 89, 119, 181
art criticism, 171
art history, 117
artificial heart, 23–24
arts, scientific approach to, 31–33
Arts and Crafts movement, 10
Ashbridge, Noel, 11
Ashmolean Museum, 71
astronomy, 60, 63–64, 130, 213, 245
astrophysics, 213
Aswan Dam, 50
Athenaeum Club, 23
Atlantic Neptune (Hollandt and Desbarres), 139
atmospheric gases, 33
atomic bomb, 23, 73
audio recording, 2, 11, 173, 213
Audubon, John James, 7, 161
Auer von Welsbach, Carl, 1, 32
Augusta, Princess (Bavarian), 80
Augusta, Princess (British), 78
Austen, Jane, 208
Australia:
  criminal colony in, 90
  telegraph lines in, 221, 223
aviation, 2, 23, 53–54, 230–32, 233, 234
axial-flow compressor, 53–54
Azzolino, Cardinal Dezio, 76

Babbage, Charles, 106–8, 130, 207, 221, 229–31

INDEX

Bach, Johann Sebastian, 143
Bache, Alexander, 108
Bache, Benjamin Franklin, 136
bacteria cultures, 142, 144
Baird, Fullerton, 19
Baker, Benjamin, 50
Balfour, Arthur, 183
Ball, Alexander, 6
ballet, 197, 231–33
balloon flights, 81–83, 136
Balsamo, Giuseppe (Cagliostro), 167–69
Balzac, Honoré de, 38, 181, 188
Bancroft, Edward, 168
Bancroft, George, 16–18
Bankes, William, 56
banking, 7, 227
Banks, Joseph, 26, 28, 88–90
Barbier de la Serre, Charles, 158
bar codes, 174
Baring, Alexander, 251
Barnum, P. T., 209, 253
Bartholdi, Frédéric, 181
Barton, Catherine, 96
Bassett, Walter, 192
Batiushkov, Konstantin, 29
Batz, Charles de, 197
Baudelaire, Charles, 181
BBC (British Broadcasting Corporation), 11, 85
Beauchamp, Pierre, 197
Beauharnais, Alexandre de, 80
Beauharnais, Eugène de, 80
Beauharnais, Josephine de, 80, 138, 229
Beaumarchais, Caron de, 81, 206
Beccaria, Cesare, 249–51
Beckford, William, 141, 149
Becquerel, Henri, 213
Beecham, Sir Thomas, 11
Beecher, Henry Ward, 190
Beethoven, Ludwig van, 59, 149, 179
Behn, Afra, 246
Bell, Alexander Graham, 240–42
Bell, Mabel Hubbard, 240
Benedict XIV, Pope, 126
Benjamin, Asher, 98
Bentley, Thomas, 235, 237
Berkeley, George, 96–98
Bernhardt, Sarah, 241–43
Bernoulli brothers, 199
Berzelius, Jons, 207–9
Bessemer, Henry, 193, 252, 254
Bethlehem Steel, 190
Bevan, E. J., 152
Beyle, Marie-Henri (Stendhal), 38
Big Bang, 64
Bigelow, John, 17–19
Bignon, Jean-Paul, 147
Biot, Jean-Baptiste, 83
Birdseye, Clarence, 12, 14
birth control, 131
births, male vs. female, 116
Bismarck, Otto von, 16
Bizet, Georges, 218
Blacas, Louis, duc de, 58
Black, Joseph, 116, 178, 217
blackband ironstone, 118–20
Blackbeard (Edward Teach), 148
Blaise Hamlet, 68
Blaxland, George, 90
blood pressure, 116
blood vessels, surgical repair of, 20–22
Bloomsbury Group, 233
blowpipes, 207–9
bluestockings, 187
Blumenbach, Johann, 117
Blumlein, Alan, 11
Boccherini, Luigi, 149
Bode, Johann, 130
Boileau, Nicolas, 86
Bolshevik Revolution, 233
Boltzman, Ludwig, 212
Bonaparte, Charles Lucien, 159–61
Bonaparte, Jérôme, 69
Bonaparte, Joseph, 159
Bonar, Charles, 209
Bonnet, Charles, 147–49
Bononcini, Giovanni, 76–78
Bonpland, Aimé, 229
Bonstetten, Charles Victor de, 177
Booth, Cecil, 192, 194
Booth, William, 51
borax, 153, 154
Born, Ignaz von, 207
Bose, Georg Mathias, 126
Boston, Mass., architecture of, 98
Boston Tea Party (1773), 155–56, 157, 225
botany, 26–28, 78, 118, 139, 148–50, 159, 200, 229, 236
Boulton, Matthew, 127, 168–70, 236
Bow Street, 165
Boyle, Robert, 247
Bragg, Gwen Todd, 223
Bragg, Lawrence, 223
Bragg, William, 223
Brahms, Johannes, 31, 93
Braille, 158
Brandenburger, Jacques, 152
Braun, Karl Ferdinand, 113
Brayley, Edward, 68
Bread and Cheese, 170
Brentano, Clemens von, 67
Bridgeman, Laura, 210
bridges, 50
Brighton Pavilion, 68
British Broadcasting Corporation (BBC), 11, 85

INDEX

British East India Company, 26, 157
British Guiana-Venezuela border, 200
British Museum, 28, 139, 161
Bronx Zoo, 70
Brooke, James, 18
Brook Farm, 188
Brougham, Henry, 109
Brown, Robert, 26–28
Browning, Elizabeth Barrett, 171, 211
Browning, Robert, 113, 171, 211
Brunel, Marc, 106
Brush, George, 30
Bryant, William Cullen, 17
bubbles, 53, 132
Buchanan, James, 17
Buffon, Georges-Louis de, 238
Bulfinch, Charles, 98
Bull Run, Union defeat at, 51
Bunsen, Robert, 83, 84
Burdett-Coutts, Angela, 18
Burke, Edmund, 228
Burlington, Richard Boyle, Lord, 76
Burne-Jones, Edward, 10, 241
Burney, Fanny, 208
Burns, Robert, 59, 189
Bush, Vannevar, 73
Bute, John Stuart, Lord, 78
Butler, Samuel, 245
Byron, George Gordon, Lord, 8, 15, 29, 56, 105, 191, 219–21, 231, 239, 249
Byron, Isabella Milbanke, 221

Cagliostro (Giuseppe Balsamo), 167–69
calculation, automated, 106, 231
calculus, 79, 126, 157
Cambridge University, 111
camera film, 102
Campani, Giovanni, 145–47
Canada:
  Franco-British conflicts over, 137–39, 196
  U.S. border with, 251
Canal du Midi, 196
Cannon, Walter, 72, 74
capitalism, 27
capital punishment, 249, 251
Capitol building, 69, 91, 98, 170–72
carbolic acid, 40–42, 153
carbon atoms, asymmetric, 33
Carême, 121–23
Caribbean plantations, 118
Carlson, Chester F., 124
Carnegie, Andrew, 91
Carnot, Sadi, 20
Caro, Heinrich, 143
Caroline (of Ansbach), Queen, 96
Caroline (of Brunswick), Queen, 175
Carothers, Wallace, 93, 94
Carpenter, Mary, 48
Carrel, Alexis, 20–22, 24
Carroll, John, 69
Carter, Howard, 50
Casanova, Giovanni Giacomo, 169
Caserio, Santo, 20
Castelnau, Julie, 181
Castries, Marshal de, 106
catalysis, 210
Catesby, Mark, 148–50
cathedrals, structure of, 103
Catherine I, Empress of Russia, 236
Catholic Church, 69, 85, 86, 87, 99
Caucasian racial type, 117
Cavendish, Henry, 116
Cayley, George, 90
cellophane, 152, 154
celluloid, 102, 104
cellulose, 152
censorship, literary, 31, 206, 243
Chambers, Ephraim, 35–36, 37
Champollion, Jean-François, 58
Channing, William Ellery, 46–48
Charles, Jules César, 136
Charles Edward (Bonnie Prince Charlie), 16, 119–21, 156, 176, 177, 248
Charles I, King of Great Britain, 77, 246
Charles II, King of Great Britain, 86, 246
Charles X, King of France, 58
Chartism, 49–51
Châtelet, Émilie du, 79, 199
Chatsworth House, 200
Chauncey, Isaac, 7
chemical symbology, 207
chemistry:
  agricultural, 63
  electron bonds in, 93
  molecular, 13
  nomenclature in, 136
  quantitative analysis in, 217
chemotherapy, 144
Chevreul, Michel, 98–100
chewing gum, 203–4
Choctaws, 60
Choiseul, Étienne-François de, 106
cholera, 42, 58
Chopin, Frédéric, 38, 40, 188, 211
Christianity:
  Catholic, 69, 85, 86, 87, 99
  Protestant, 8, 18, 86, 87, 99, 148, 178, 226
  reforms of, 96, 218
Christian Socialism, 18
Christina, Queen of Sweden, 76
chromosomes, 222
Church of England, 8, 18

Church of Scotland, 178
Cierva, Juan de la, 230–32, 234
citizen armies, 130
Civil War, 17, 51–53, 60, 70, 91, 110
Clafin, Tennessee, 190
Clairmont, Claire, 219–21
Clarke, Edward, 27
Clarke, Samuel, 96
Claude, George, 244
Clayton, Thomas, 87
Clement, Joseph, 106
Clementi, Muzio, 29
Cleopatra's Needle, 50
Clerisseau, Charles-Louis, 176
Clerk, John, 216
clingwrap, 134
Clinton, George, 137
clocks, 184, 247
coal mines, 196
coal tar, 120, 123, 143, 153
Coast Guard, U.S., 225
Cobbett, William, 198
Cochrane, Thomas, 227
Cockerell, Christopher, 194
coherer, 173
coins, copper, 96
coin-stamping machines, 127
Cole, William, 126–28
Coleman, William Tell, 153
Coleridge, Samuel Taylor, 6, 15, 47, 219, 239
collodion, 100–102
Colony Aid Association, 110
colors, juxtaposition of, 100
Colson, John, 126
Columbus, Christopher, 1
Combe, George, 208–10
combustion, effects of atmospheric pressure on, 163
comic opera, 123
composers, 41–43, 86, 93, 99, 146–48, 149
  of opera, 76–78, 87, 99–101, 123, 169, 179, 205, 207, 230, 236–38
compressor blades, 53–54
Compton, Henry, 148
computers, 221, 231
Condorcet, Marquis de, 157–59
Conduitt, John, 96
Confederate States of America, 110
Congreve, Sir William (artillerist), 227
Congreve, William (dramatist), 96, 116
Constable, John, 209–11
Constant, Benjamin, 177, 178
Constitution, U.S., 206
Constitution, USS, 5, 6, 7
contact lenses, 163–64
contagion theories, 58
Cook, James, 47, 67, 90, 139, 141, 207
Cooke, W. F., 221
Cooper, James Fenimore, 170, 179
copper money, 96
Cornell, Ezra, 133, 138
Cornell University, 131–33
Cornwallis, Lord Charles, 78, 216–18
Coronation of Napoleon (David), 138
Corsican revolution, 156
Cosway, Maria, 91
counterfeit money, 227
Covent Garden Theater, 198
cowpox, 56
creation:
  chain of being at, 149
  date of, 129
Creation, The (Haydn), 59
creosote, 123, 151
Crick, Francis, 224
Crimean War, 160, 220
criminology, 249–51
Crockett, Davy, 203
Crookes, Sir William, 113, 130–32, 173, 213
crop rotation, 248
Cross, Charles, 152
Crozat, Antoine, 88
Crozat, Pierre, 88
crystal-diffraction studies, 2, 83
Crystal Palace Exhibition, 200, 202
crystals:
  liquid, 182, 184
  measurement of, 191
Cuffe, Paul, 186
Cullen, William, 178, 217
Curie, Marie, 213
Custis, George Washington, 91
cybernetics, 74
Cyclopedia (Chambers), 35–36, 37

Daguerre, Louis-Jacques-Mandé, 138
Dallos, Joseph, 162, 164
Dalrymple, John, 88
Dalziel, Andrew, 178
dance, 197, 231–33
da Ponte, Lorenzo, 169
d'Arblay, Alexandre, 208
d'Arsonval, Arsene, 242, 244
Daru, Pierre, 38
Darwin, Charles, 18, 20, 111, 129, 161–63, 210, 216
Darwin, Erasmus, 236
David, Jean-Louis, 138
Davis, Jefferson, 110
Davy, Humphry, 37, 98, 130, 189
Deane, Silas, 168, 206
Debussy, Claude, 181
Declaration of Independence (1776), 217
Deffand, Marie de Vichy-Chamrond du, 128

INDEX

Defoe, Daniel, 248
De Forest, Lee, 173, 174
Degotti, Ignazio, 138
Delacroix, Eugene, 38, 188, 211
Delaware, Lackawanna, and Western Railroad, 140
democracy, 178, 181–83, 220
De Morgan, William Frend, 109–11
Denny, William, 51
Desbarres, Joseph, 139
Descartes, René, 1, 245
Des Cloizeaux, Alfred-Louis-Oliver Legrand, 83
de Valera, Eamonn, 220
Devonshire, William Cavendish, Duke of, 198–200
Dewar, James, 132, 134, 143
Diaghilev, Sergey, 233
dialectical materialism, 49
dictionaries, 18, 37
Diderot, Denis, 36, 89
Didot St. Leger, 136
didymium, 30, 32
dietary deficiency, 42, 43
difference engine, 106, 231
digestion, 72, 217
digital clocks, 184
dioramas, 68, 138
Disney, Walt, 11, 67
DNA, 223–24
Dobereiner, Johann Wolfgang, 151
Douglas, John, 89
Drummond, Thomas, 39
Drury Lane Theater, 66, 87, 99, 165, 167, 197, 208, 228
Dryden, John, 86
Drysdale, John, 178
Drysdale, Mary Adam, 178
Dupin, Amadine-Aurore-Lucille (George Sand), 38, 171, 188, 211
DuPont Company, 93
Durand, Asher, 170
Dvorak, Antonin, 93
dyes, 123, 143, 144, 218

Earth:
  age of, 129, 216
  shape of, 201, 247
Eastman, George, 102
Eckersley, Peter, 9–11
Eckersley, Stella, 9
ecology, 210
economics, 36, 157, 180, 231
Eddystone lighthouse, 127
Edgerton, Harold, 33
Edinburgh Oyster Club, 215–16, 217
Edinburgh Review, 109
Edison, Thomas, 1, 72, 240
Edison effect, 173
education reforms, 48, 129–31, 186, 236, 238
Edward VII, King of Great Britain, 192, 241
Egerton, Francis, 196
Egyptian antiquities, 50, 56–58, 71
Ehrenberg, Christian, 28–30
Ehrlich, Paul, 144
Eiffel, Gustave, 181
Eijkman, Christiaan, 42, 44
Eilshemius, Fanny Angelina, 140–42, 144
Eilshemius, Louis, 140
Einstein, Albert, 220
Eire, Republic of, 220
electricity, storage of, 125–26, 127
electric light, 1, 30
electromagnetism, 100, 113, 130, 180
electron bonds, 93
electronics industry, 213
"Elegy in a Country Courtyard" (Gray), 177
elements, discoveries of, 30, 33, 136, 191, 198, 207
elephants, extinction threat to, 100
elevator, 253–54
Elgin, Thomas, Lord, 57–59
Ellis, Henry, 166–68
Elster, Julius, 112
Ely, Richard, 180
Emperor fountain, 200
encyclopedias, 35–36, 89, 109, 147, 201
energetics, 212
Engels, Friedrich, 49
engine knock, 13
Enlightenment, 6–8, 36, 89, 115, 157, 178, 201, 215
espionage, 119, 168, 176, 248
Essay on Criticism, An (Pope), 116
Essay on Tactics (Guibert), 128–30
ether, nonexistence of, 180
ether spray, 40
Eugénie, Empress of the French, 90, 218–20
Euripides, 239
Everett, Edward, 251–53
Everett, Percival, 202, 204
evolution, 163, 173, 216
extraterrestrial life, 245–46, 247
extruder, 133–34

F-117A (stealth fighter), 2, 214
Fabre, Francois-Xavier, 121
Fairbairn, William, 240
Falklands War (1770), 105–6, 107
Faneuil Hall, 98
Fantasia, 11
Faraday, Michael, 130

INDEX

Farmer, George, 107
Farragut, David, 17, 60
fashion industry, 90–92, 123, 218
fats, chemistry of, 98–100
Federalists, 136
feedback systems, 72, 73–74
feral child, 115–16
Ferdinand II, Grand Duke of Tuscany, 146
Ferris Wheel, 190–92
fertilizer, 63, 229
fêtes galantes, 86
Fick, Adolf, 163, 164
Field, John, 29–31
Field, Kate, 241
film cartridges, 102
films, sound in, 173
Firmian, Leopold von, 99
fish, frozen, 12, 14
Flaubert, Gustave, 31, 181, 188, 243
Flaxman, John, 237
Follen, Karl, 61
Folsom, Charles, 17
Fontenelle, Bernard de, 246
Fonthill Abbey, 141, 149
food insulation material, 134
food supply, population growth vs., 27
forgery, protection against, 227
Forster, George, 67, 207
Forth Bridge, 50
fossils, 146
Fouquet, Nicholas, 197
Fowler, Sir John, 50
Fox, Charles James, 228
France, 1848 revolution in, 49, 58
Francis II, Holy Roman Emperor, 230
Frankland, Edward, 163
Franklin, Benjamin, 38, 57, 88, 108, 148, 168, 187, 198, 217
Franklin, Rosalind, 223, 224
Fraunhofer, Joseph von, 82, 84
Frederick, Prince, 78
Frederick II (the Great), King of Prussia, 130, 149, 199
Freemasonry, 167–69, 207
French Revolution, 80, 81, 136, 137, 157–59, 198, 206, 208
Freon, 13, 14
Friedrich Wilhelm II, King of Prussia, 149
frigidity, effects of, 132
frozen fish, 12, 14
Fuller, Margaret, 171
Fulton, Robert, 170, 250
Furnivall, Fred, 18–20

galaxies, retreating, 62, 64
Gall, Franz Joseph, 208
gallium, discovery of, 30
Galton, Francis, 111
Galvani, Luigi, 118
Garcia, Manuel, 40
Garibaldi, Giuseppe, 50, 110
Garrick, David, 66, 166, 167, 208, 226
Garstin, William, 50
gases, atmospheric, 33
gaslights, 1, 30–31, 123, 143, 151, 163
gas liquefiers, 242, 244
gasoline, 13, 242
gastronomy, 121–23, 209
gasworks, 120, 123, 143
Gauss, Friedrich, 28
Gay, John, 248
Gay-Lussac, Joseph-Louis, 83, 130
Geitel, Hand, 112
gelatin, 142
Genêt, Edmond-Charles-Edouard, 137
geology, 37, 129, 146, 191, 216, 238, 240
geomagnetism, 28, 108
George II, King of Great Britain, 78, 89
George III, King of Great Britain, 57, 78, 130, 156
George IV, King of Great Britain, 109
Georgia, founding of, 97–99
German nationalism, 61, 230
Gilbert, W. S., 41
Gillot, Claude, 86
glacier flow, 163, 240
Gloster E29, 54
gloves, surgical, 120–22
Gluck, Christoph, 236–38
Goddard, Robert, 23
Godwin, William, 47, 219
Goethe, Johann Wolfgang von, 6, 15, 16, 31, 100, 151, 177, 178, 179, 201, 249
Good, John Mason, 158
Goodyear, Charles, 120
Gothic novels, 128, 141, 176
Gothic Revival, 171–73
government borrowing, 96, 248
Graham, George, 247
Grand Tour, 76, 176–77, 178, 196
Grant, Anne, 59–61
Grasse, François-Joseph-Paul de, 118, 216–18
Gray, Matthew, 133, 134
Gray, Thomas, 177
Great Awakening, 226
Great Chain of Being, 149
Great Nosology Book, 217
Great Southern Continent, 139
Great Weston Meteor, 37
Greek myths, 117

INDEX

Greek scholars, 237–39
Greeley, Horace, 169–71
Greenhow, Rose, 51
Greenland seas, 75
Greenwich Observatory, 221–23
Gregory, Olinthus, 158
Grenfell, Wilfred, 10
Griffith, A. A. (inventor), 53, 54
Griffith, Arthur (journalist), 220
Grimm brothers, 67–69
Groot, Hugo de, 76, 77
Grouchy, Sophie de, 157
ground effect, 193
Grove, George, 9
Guibert, Joseph de, 128–30
Guizot, François, 49
Gully, John, 111
gun cotton, 100, 102
gunpowder, 159, 160
gun technology, 118, 168
gutta-percha, 21, 70, 133
Gwynn, Nell, 246

Hadfield, George, 91
Haeckel, Ernst, 210
hairdressing, 153, 231
Hale, George, 213
Hales, Stephen, 116–18, 166
Halifax, Charles Montagu, Lord, 96
Hall, Basil, 39
Hall, Spencer Timothy, 61–63
Halsted, William, 120
Hamilton, Lady Emma, 59, 228
Hamilton, Sir William, 59, 109–11, 228, 237, 239
Handel, George Frideric, 76–78
Hardy, Thomas (author), 41
Hardy, Thomas Masterman (navy captain), 39–41
Hare, Robert, 39
Hare-Naylor, Georgiana, 237
Harford, John Scandrett, 66–68
Harley, Robert, 248
Harris, Renatus, 148
Harvard College, 61
Hastings, Battle of (1066), 2
Hastings, Warren, 228
Hatshepsut, mortuary temple of, 71
Hauy, R. J., 158
Hauy, Valentine, 158
Haydn, Franz Joseph, 29, 59, 149, 207
Hayes, Catherine, 40
Hazlitt, William, 239
Hearn, Lafcadio, 60
heart surgery, 23, 24
Hegel, G. W. F., 180
Heine, Heinrich, 6
helicopters, 234
hemlines, 92, 94
Henley, W. E., 153
Henry, Joseph, 138
Henry, Prince, 156
Hensen, Victor, 210
Herapath, William Bird, 103
herapathite, 103, 104
Herder, J. G., 15
Herschel, Caroline, 130
Herschel, William, 8, 130
Hertz, Heinrich, 180–82
Hesse, Fanny Angelina Eilshemius, 140–42, 144
Hesse, Walther, 140–42
Heyne, Christian Gottlob, 117
hieroglyphics, 58
high-definition television, 11
Hill, Matthew, 48
Hill, Rowland, 48
Hippisley, John Coxe, 156–58
history, dialectical process of, 180
history painting, 57, 138, 170–72, 250
Hitler, Adolf, 230
Hoffman, Charles Fenno, 169–71
Hoffmann, August, 143
Hoffmann, Ernst Theodor, 81
Hogarth, David, 71
Holland, Elizabeth Webster, 121
Holland House, 121
Hollandt, Samuel, 139
Holley, Alex, 193
Holmes, Oliver Wendell, 58–60
Holyoake, George, 110
Home, D. D., 111–13
Home Insurance Company building, 254
homeostasis, 72, 74
Hooke, Robert, 146, 247
Hopkins, Gowland, 43, 44
Hopkins, William, 240
Horsfield, Thomas, 26–28
hot-air balloons, 81–83
hovercraft, 194
Howe, Julia Ward, 210
Howe, Samuel, 210
Hubbard, Gardiner, 240–42
Hubble, Edwin, 64
Hudson River School, 170
Hugel, Charles von, 90
Humboldt, Alexander von, 28, 30, 130, 207, 229
Humboldt, Wilhelm von, 6, 16
Hunt, (William) Holman, 48, 239
Hutchinson, Tom, 156
Hutton, James, 129, 178, 216
Huxley, Thomas Henry, 161–63
Huyghens, Christiaan, 146, 247
Hyatt, John Wesley, 100–102
hydrogen, 136
hymns, Methodist, 226

INDEX

ice patrols, 53
Illinois Central Railroad, 70
Impressionist painting, 100
Industrial Revolution, 1, 10, 27, 49, 120, 168, 250, 251
inert gases, atmospheric, 33
Ingres, Jean-Auguste-Dominique, 38
insurance industry, 187
Intrepid, USS, 5
investors' information service, 191
iodine, antiseptic use of, 160
Ireland, William, 101
Irish independence, 51, 150, 220
iron furnaces, 83, 120
ironstone, 118–20
Irving, Washington, 16, 71, 201
Isouard, Nicolas, 123
Italy, unification of, 110, 171

James, Henry, 21–23, 48
James Edward, Prince, 88, 119, 156, 176
James I, King of Great Britain, 77
James II, King of Great Britain, 86, 87, 148
Jameson, Robert, 129
Jefferson, Thomas, 61, 89–91, 141, 159, 176
Jeffreys, George, 148
Jekyll, Elizabeth Somers, 97
Jekyll, Sir Joseph, 97
Jenkin, Fleeming, 21
Jenner, Edward, 56
Jenney, William, 254
Jesuits, 69
jet airplanes, 54
Johnson, Samuel, 37, 66, 141, 166, 228
Jones, William (American financial official), 7
Jones, William (British Sanskrit translator), 66, 67
Journal des Savants, 147
journalism, 87, 135–36, 137, 160, 171, 186, 197–99
Juniper Hall, 45–46, 47
juvenile delinquency, 48

Karl August, Duke of Weimar, 151
Kato, Yagoro, 213, 214
Keats, George, 7
Keats, John, 7
kelp industry, 160
Kemble, John, 198
Kent, James, 170
Kent, William, 76, 78
Kew Royal Botanic Gardens, 78, 90, 152
Key, Francis Scott, 227
Keynes, John Maynard, 27, 231, 233
Kidd, William, 97
King, William Rufus, 17
Kinglake, William, 160
Kingsley, Charles, 18, 20, 171
Kirchoff, Gustav, 84
Kit-Kat Club, 87, 95–96, 98, 199
Kleist, Ewald von, 126
Kneller, Godfrey, 87–89
Knox, William, 168
Koch, Robert, 42, 142, 144
Kodak cameras, 102
Koenig, Samuel, 199–201
Komensky, Jan Amos, 77–79
Kôsciuszko, Tadeusz, 36–38
Kossuth, Lajos, 220

labor, division of, 1, 157, 215
Labrador Mission, 10–12
Lacepède, Bernard, 238
Lafayette, Marquis de, 159, 170, 188, 206–8
laissez-faire economics, 36
Lake Poets, 47
Lamartine, Alphonse de, 136, 158, 160, 181, 249
Lamb, Charles, 47, 219, 239
Lampton, Curtis Miranda, 70
Land, Edwin, 104
Landor, Walter Savage, 239–41
landscape painting, 140, 170, 209–11
Langmuir, Irving, 13
language, universal, 77–79, 188, 210
Laplace, Pierre-Simon, 130
Lardner, Dionysus, 8
la Rochefoucauld, François, Duc de, 159
laryngoscope, 40
lasers, 1, 34
Latrobe, Benjamin Henry, 69–71
Latrobe, Charles, 71
Lauder, William, 89
Lavoisier, Antoine, 136–38, 198
Law, John, 88
Lawrence, T. E., 71–73
Lazarus, Emma, 183
lead poisoning, 13
Leavitt, Henrietta, 63, 64
Le Breton, Andre, 36
Lecoq de Boisbaudran, Paul Émile, 30–32
Lee, Arthur, 206
Lehmann, Otto, 182
Leibniz, Gottfried, 79
lens-crafting, 80–82, 145, 147
Leroy, Achille (St. Arnaud), 160
Lespinasse, Julie de, 128, 201
Lesseps, Ferdinand de, 218
Lesueur, Charles-Alexandre, 238
Lewis, Gilbert, 93
lexicons, 37

Leyden jar, 125–26, 127
Liberia, establishment of, 186
Liebig, Justus von, 30, 63
life:
  DNA encryption of, 222, 223–24
  extraterrestrial, 245–46, 247
life insurance, 187
light, polarization of, 83, 103, 104
Light Brigade, charge of, 160
lighting, 1, 30–32, 39, 163, 244
light-scattering effect, 183, 184
"Lillibulero," 85–86, 87
limelight, 39
Lincoln, Abraham, 171
Lind, Jenny, 40, 209, 221
Lindbergh, Charles, 23–24
Linley, Thomas, 99–101, 228
Linnaeus, Carolus, 139, 150, 166
liquid crystals, 182, 184
Lister, Joseph, 40–42, 153
Liszt, Franz, 38
Literary Club, 37, 66
literary frauds, 15–16, 89, 101
Livingston, Robert R., 170, 250
Lodge, Oliver, 173
logical atomism, 231
Lohmeyer, Ferdinand, 93
London Diorama, 68
London subway system, 50
London Zoo, 26
Lopukhova, Lydia, 231–33
Louisiana Territory, 88, 137, 251
Louis XIII, King of France, 77, 197
Louis XIV, King of France, 86, 185, 195, 196, 197
Louis XV, King of France, 36, 46, 86
Louis XVI, King of France, 46, 81, 206, 208
Louis XVIII, King of France, 58
Louis-Philippe, King of the French, 49
Loutherbourg, Philippe Jacques de, 167
Lovelace, Lady Ada Byron, 221
Lowe, Sir Hudson, 150
Lowell, Percival, 60
Ludwig II, King of Bavaria, 230
Lunar Society, 196, 236
Lundy, Benjamin, 186
Lusieri, Giovanni Battista, 57–59
Luti, Benedetto, 76
lycopodium, 122, 124

Maastricht, siege of, 195–96, 197
McClellan, George, 51, 108
MacGillivray, John, 161
MacGillivray, William, 161
Macintosh, Charles, 8, 120
McKenzie, Henry, 189
Macklin, Charles, 166
Maclure, William, 129, 131, 238
MacPherson, James, 15–16, 189
Madame Bovary (Flaubert), 31
Magic Flute, The (Mozart), 207
Magie, W. F., 70–72
magnetic poles, 28, 108
magnetic variation figures, 108
Malthus, Robert, 27
Man in the Iron Mask, 185
Mann, Horace, 176
Manzoni, Alessandro, 249
Marconi, Guglielmo, 11
Marie-Antoinette, 81, 167
marine biology, 210, 238
Marriage of Figaro, The, 81, 205–6, 207
Mars, 23, 60
Marsh, Othniel, 30
Martineau, Harriet, 61
Marx, Karl, 29, 49, 180
Mascagni, Paolo, 150
Masséna, André, 38
Massenet, Jules, 15
mathematicians, 116, 126, 157, 199, 212, 229–31
Maudslay, Henry, 106
Maupertuis, Pierre de, 199–201, 247–49
Maury, Matthew Fontaine, 51–53
mauve, 123, 143
Maximilian I, King of Bavaria, 80
Maybrick, Florence, 43
Maybrick, James, 43
Maybrick, Michael (Steven Adams), 41–43
Meade, Richard, 89
Mecklenburg-Schwerin, Grand Duke of, 141–43
medical training, 217
Medieval English, 101
Mediterranean shipping, 5
Meigs, Montgomery, 91
Melville, Henry Dundas, Lord, 107–9
Melville, Robert, 118
Mendelssohn, Felix, 6, 15, 31
Meredith, George, 41
Mérimée, Prosper, 8, 188
Mesmer, Franz, 81
metal, stress studies on, 53
meteorites, 37–39
Methodism, 226
methylene blue, 143, 144
metric system, 231
Metternich, Clemens von, 90, 230
Metternich, Pauline von, 90
Michelson, Albert, 180
micropaleontology, 30
Midgley, Thomas, 13, 14
military:
  field hospitals of, 88
  nonprofessional, 128–30

INDEX

training program in, 46
trench warfare in, 195–96
Mill, John Stuart, 231
Millais, John, 48–50, 241
Milton, John, 89
mineralogy, 30
miner's safety lamp, 98
Minot, Charles, 140
Mission San Antonio de Valero (Alamo), 203
Monckton, Robert, 248–50
Mond, Robert Ludwig, 50–52, 54
money-back guarantee, 235
Monist League, 210
Monroe Doctrine, 200
Montagu, Elizabeth, 139–41, 187
Montagu, Mary Pierpont Wortley, Lady, 199
Montgolfier brothers, 81
Montijo, Countess of, 218
Montreal, 1760 siege of, 137–39
Moody, Dwight, 10
Moore, C. C., 169
More, Hannah, 66, 186, 187
Morellet, André, 89
morphology, 100
Morris, Gouverneur, 206
Morris, William, 10
Morse, Samuel, 133, 138, 172, 174, 221
Mouries, Meges, 98–100
Mozart, Leopold, 99
Mozart, Wolfgang Amadeus, 81, 99, 149, 169, 205–6, 207
Muller, Johan von, 177
munitions, 118, 136, 143, 168, 227
Mushet, David, 118
Musschenbroek, Petrus van, 127
Musset, Alfred de, 38–40, 188
myths, scientific approach to, 117

Nadar (Gaspard-Félix Tournachon), 243
naphtha, 120, 123
Napier, Macvey, 109
Napier, Robert, 9
Napoleon Bonaparte, 6, 15, 29, 58, 121, 131, 156, 178, 201
brothers of, 59, 169
coronation painting of, 138
death of, 249
defeats of, 25, 80, 107, 239
in exile, 39, 150, 159
military conquests of, 38, 67, 80, 130, 149, 177, 218
scientific interests of, 98, 238
son of, 80
Napoleon III, Emperor of the French, 160, 218, 220
Narbonne, Count of, 46
Nash, John, 68
Nashoba commune, 186, 188
National Conservatory of Music, 93
National Vaccine Institution, 56
Native American culture, 60, 236
naturalism, 20
natural law, 36
Naturphilosophie, 100, 180
Naval Academy, U.S., 53
naval warfare, 7, 25–26, 60, 107, 190, 193, 216–18, 227
Naville, H. E., 71
Naville, Marguerite, 71
Nazism, 67, 111, 210, 220
Necker, Jacques, 177, 201, 206
Necker, Suzanne Curchod, 201
Neef, Joseph, 131, 238
Neilson, James Beaumont, 120
Nelson, Lord Horatio, 6, 25, 26, 27, 39–41, 59, 107, 216, 228, 239
neoclassicism, 57
neodymium, 1, 32, 34
neon lights, 244
Nernst, Emma Lohmeyer, 93
Nernst, Walter, 93
Nestlé, Charles, 154
neural transmitters, 72
Newburyport, Mass., 225–26, 227
New Christianity, 218
Newcomb, Simon, 180
New Harmony, Ind., utopian community at, 131, 191, 238
newspapers, 135–36, 137, 160, 171, 186
Newton, Sir Isaac, 79, 89, 96, 111, 126, 146, 247
New World Symphony (Dvorak), 93
New York, governorship of, 137
nickel alloys, 52, 54
Nile, Battle of (1799), 27, 107
Nimonic 80, 52, 54
Nobel prizes, 33, 44, 93, 144, 183, 210, 224
Northern Star, 49–51
Noyes, Arthur, 213
nylon, 94

obstetrics, 58
O'Connell, Daniel, 150
O'Connor, Feargus, 51
Oersted, Hans Christian, 207
Office of Scientific Research and Development, 73
offshore limit, 77
Oglethorpe, James, 97–99
ohm, 21
O'Meara, Barry, 150
Open Seas, 76, 77

INDEX

opera, 40, 76–78, 87, 99–101, 123, 169, 179, 205, 207, 209, 230, 236–38
ophthalmology, 162, 163–64
organs, construction of, 148
organ transplants, 24
ornithology, 161
Orton, William, 240
Osborn, Henry, 70
Osborn, William Henry, 70
Osbourne, Fanny, 153
oscilloscope, 113, 114
osmotic pressure, 33
Ossian, 15–16, 17, 189
Ostwalt, Wilhelm, 210–12
Oswald, Richard, 187, 189
Otis, Elisha, 253, 254
Owen, David Dale, 191
Owen, Robert (father), 131, 191
Owen, Robert Dale (son), 131, 191
Oxford English Dictionary, 18
oxygen, 136, 198
oxyhydrogen blowpipe, 39
ozone hole, 13

Paganini, Niccolò, 31
Paine, Thomas, 217–19
painting:
  color theory in, 100
  of historical subjects, 57, 138, 170–72, 250
  of landscapes, 140, 170, 209–11
  of portraits, 57, 87, 138, 141, 170, 226–28, 241
  Pre-Raphaelite style of, 10, 48–50
paleontology, 30, 146
Palestine Exploration Fund, 9
Palladio, Andrea, 76
palladium, 191
Pan Am transatlantic service, 233
pansophy, 77, 79
Paoli, Pascal, 156
papal states, politics of, 110
papermaking machine, 136
paper money, forgery of, 227
Paradies, Domenico, 99
Paris Diorama, 138
Parkinson, William Standway, 27
Parthenon, 57
Pasco, John, 26
Patterson, Betsy, 69
Paulding, James K., 203
Paxton, Joseph, 200
Payen, Alselm, 152
Peabody, George, 30, 70
Pedro I, Emperor of Brazil, 227–29
Peel, Robert, 251
Penny Black, 48
percussion and auscultation, 58
Perkin, William, 123, 143
Perkins, Jacob, 227
permanent-wave styling, 154
Perry, James, 239
Pestalozzi, Heinrich, 129–31, 238
Petrie, Flinders, 9, 50
Philadelphia General Advertiser, 135–36, 137
Philadelphia Orchestra, 11, 13
Phillips, Molesworth, 47
Phillips, Susanna, 47
photocopying, 124
photoelectric cell, 112, 114
photography, 68–70, 102, 138, 243
photometer, 63
photosynthesis, 149
phrenology, 61–63, 203, 208–10
physiotherapy, 242
pianists, 29–31
Pierce, Franklin, 46, 108
Pinkerton, Allan, 51
piracy, 97, 148
Pius IX, Pope, 110
Place, Muesnier de la, 136
planets, distances between, 130
plastics, 134
Playfair, John, 127–29
Playfair, Sir Lyon, 200–202
Playfair, William, 127
Pluto, 60
Poe, Edgar Allan, 46, 179–81
poetry, neoclassical rules of, 86
Poincaré, Jules-Henri, 213
Poiret, Paul, 92
Polaroid sunglasses, 104
police, 165, 251
political parties, 135, 171
polymers, 93–94
Pompadour, Madame de (Jeanne-Antoinette Poisson), 36, 106
Poor, Henry Varnum, 191
Pope, Alexander, 89, 116, 199
Popoff, Admiral, 193
population growth, 1, 27
porcelain, 36
Porson, Richard, 239
Portales, Albert du, 71
portrait painting, 57, 87, 138, 141, 170, 226–28, 241
postage stamps, 48, 143
postal service, 48
postcards, 202
potato blight, 150–52, 202
pottery, 196, 236, 237
Pouillet, C. S. M., 47–49
Preble, Edward, 6
predicates, 109
Preparation 606 (Salvarsan), 144
Pre-Raphaelite Brotherhood, 10, 48–50, 241
press-ganging, 97

INDEX

Price, Richard, 187
Priestley, Joseph, 37, 88, 116, 198, 236
Pringle, John, 88
prison reform, 48, 156–58, 249
Proclamation Society against Vice and Immorality, 37
protein, muscular energy from, 163
Protestantism, 8, 18, 86, 87, 99, 148, 178, 226
psychic phenomena, 39, 111–13, 132, 167, 173
public housing, 68, 70
puerperal fever, 58
Puffing Billy, 192, 194
Pugin, Augustus, 68
pulley blocks, machine-made, 106
Purcell, Henry, 86, 146–48
Puritans, 77, 246
Pushkin, Aleksandr, 29
Pyramidology, 9

quantitative analysis, 217
Queensware, 236
Quesnay, François, 36
Quetelet, Lambert, 111
quinine, iodosulfate of, 103

Rachmaninoff, Sergey, 181, 233
racial types, 117
radar, 214
radiative scattering, 183, 184, 212, 214, 223
radio, 11, 73
radioactivity, 213
radium radiation, 132
Raevsky, Nickolay, 29
Raffles, Stamford, 26
Raglan, Fitzroy Somerset, Lord, 160
railroads, 70, 91, 108–10, 138, 140, 151, 171, 191–93, 221
raincoats, 120
Ramée, Joseph, 141
Ramsay, William, 33
rare earths, 30–31
Rayleigh, Evelyn Balfour Strutt, Lady, 183
Rayleigh, John William Strutt, Lord, 183, 184
Reade, J. Bancroft, 68–70
realism, literary, 31
Reaumur, René-Antoine de, 147
recording technology, 2, 11, 173, 213
reductionism, 1, 2, 31
reflector telescopes, 63
refrigeration, 13, 14
Reinitzer, Friedrich, 182, 184
religious sermons, 226
reproductive physiology, 81
Republican Party, 135, 171
respiratory medicine, 118
Revolutions of 1848, 49, 58, 220
Reynolds, Joshua, 141, 166, 208
Rich, John, 248
Richardson, Benjamin, 40
Richelieu, Armand-Jean du Plessis, Duc de, 77, 79
Rights of Man, The (Paine), 219
Rimsky-Korsakov, Aleksandr, 38
Ripley, George, 188
Robert, Nicholas, 136
Robinson, John, 250–52
Rochambeau, Comte de, 78–80
rocketry, 23, 227
rocks, origins of, 129, 216, 240
Rodney, George Brydges, 118, 216
Roebuck, John, 118
Roentgen, Wilhelm Conrad, 213
Rogers, Henry, 238–40
Rogers, Samuel, 8–10
Roman antiquities, 176, 237
Romantic movement, 6, 7, 8, 15, 47, 49, 67, 100, 151, 176, 178–80, 189, 236
Romilly, Sam, 251
Romney, George, 226–28
Roscher, Wilhelm, 180
Rose, George, 56
Rose, William, 56
Rosenblueth, Arturo, 72, 74
Rosetta Stone, 58
Ross, James Clark, 28
Rossetti, Dante Gabriel, 10, 48, 241
Rossini, Gioacchino, 38, 121, 123, 179, 243
Rougon-Macquart Family, The (Zola), 20
Rouquette, Père Adrien, 60
Rousseau, Jean Jacques, 236, 239
Royal Academy, 57, 141, 167, 241, 250
Royal Institute for Blind Children, 158
Royal Society, 23, 88, 108, 146, 161
rubber, vulcanization of, 120
Runge, Frederich, 151–53
Runge, Philipp, 100
Rush, Benjamin, 217
Ruskin, John, 171–73, 241
Russell, Bertrand, 231
Russell, Henry, 179
Russell, John Scott, 252
Russell, William, 160
Russia, Alaska sold by, 19

Sabatier, Apollonie, 181
Sabine, Edward, 108
saccharimeter, 83
Said, Mohammed, Khedive of Egypt, 218
St. Arnaud (Achille Leroy), 160

saints, canonization of, 126
Saint Simon, Henri de, 218
St. Vincent, John Jervis, Lord, 107
Sakuntala, 66, 67
Salieri, Antonio, 169
Sallo, Denys de, 147
Salvarsan (Preparation 606), 144
Salvation Army, 51
Sand, George (Amadine-Aurore-Lucille Dupin), 38, 171, 188, 211
Sandeau, Jules, 188
Sanskrit, 65, 66, 67
Santa Anna, Antonio López de, 203
Santa Catharina, 75
sap, 116, 118, 166
Sarawak, white rajahs of, 18
Sargent, Epes, 179
Saunders-Roe N1, 194
Schelling, Friedrich von, 100, 178–80
Schiller, Friedrich, 177–79, 201
Schlegel, August von, 178
Schlegel, Friedrich von, 67
Schleiermacher, Friedrich, 6–8
Schoenberg, Isaac, 11
Schomborgk, Robert, 200
Schonbein, Christian, 100
Schrodinger, Erwin, 220–22, 224
Schroeder, Bill, 214
Schumann, Clara Wieck, 31
Schumann, Robert, 31
science fiction, 243
science journals, 147
Scott, Sir Walter, 8, 16, 21, 56, 59, 189, 239
Scott, Thomas, 91
Scott, Winfield, 46
Scottish culture, 16, 59, 189
Scottish Enlightenment, 178, 215
Scriblerus Club, 116, 248
sea-trade policies, 75–76, 77
Second Bank of the United States, 7
Secret Service, 51
secularism, 110
Selden, John, 77
Selma, Ala., 17
Semmelweis, Ignaz, 58
Sèvres porcelain, 36
Seward, William, 19
Shadwell, Thomas, 146
Shakespeare, William, 101, 166
Shaw, F. G., 188
Shaw, George Bernard, 10
Shelburne, William Petty, Lord, 187
Shelley, Mary, 8, 219, 221
Shelley, Percy Bysshe, 8, 219, 239
Shepherd, Thomas Hosner, 68
Sheridan, Elizabeth Ann Linley, 101, 228
Sheridan, Richard Brinsley, 8, 37, 101, 228–30
Shipley, William, 118
Short, William, 159
Siddons, Sarah, 198, 208, 210
Sieberling, Frank, 120
Sikorsky, Igor, 233, 234
silk stockings, 92, 94
Silliman, Benjamin, 37–39
Silver, Bernard, 174
singers, 40, 179, 209
Sinn Fein, 220
Sismondi, Simonde de, 8, 27–29
Sistine Chapel, 66
skiascope, 70–72
Skoda, Joseph, 58
sky, blue color of, 183
skyscrapers, 254
slavery, 37, 61, 110, 171, 186
Slipher, Vesto, 60–62, 64
smallpox, 55–56, 57, 89, 199
Smeaton, John, 127
Smibert, John, 98
Smith, Adam, 1, 15, 89, 129, 157, 178, 215, 250
Smith, Bernard, 148
Smith, Hamilton, 191
Smith, James Edward, 150
Smith, J. Lawrence, 30
Smithson, James, 131, 191
Smithsonian Institution, 131, 191
Smyth, Charles Piazzi, 9
soap bubbles, 53
socialism, 10, 27–29, 131, 180, 188, 238
social mathematics, 157
Society for Psychic Research, 173
Society of Arts, 118
Solander, Daniel, 139
solar system, 245
Somers, Lord John, 97
Somerville, Mary, 130, 221
Sommerfeld, Arnold, 212, 214
sound:
  in films, 173
  speed of, 158
  stereophonic, 11
southern magnetic pole, 28
Southey, Robert, 47
spaceflight, 23
Spallanzani, Lazzaro, 79–81
spectrograph, 60–62
spectroscopy, 30, 82, 84
spiritoscope, 39
spiritualist mediums, 39, 111–13, 167, 173, 190, 213
Spottiswood, Alexander, 148
Spurzheim, Johann Christoph, 208
Staël, Germaine de, 6, 46, 177, 178, 201
stage design, 167

stamps, adhesive, 48
Standard & Poor's 500, 191
Stanford, Edward, 160
Stanhope, Lady Hester, 158–60
stars, distance of, 63–64
statistics, 111, 116
Statue of Liberty, 181–83
stealth aircraft (F-117A), 2, 214
steamboats, 170, 193, 250
steam engine, 118, 127, 168–70, 217
Steele, Richard, 87, 116, 197, 199
steel technology, 190–92, 193, 252, 254
Steinheil, August, 63
Stendhal (Marie-Henri Beyle), 38
Stensen, Niels, 146
Stephenson, George, 98
Stephenson, Robert, 9
stereochemistry, 33
stereo recording, 11
Stevens, Isaac Ingalls, 108
Stevenson, Robert Louis, 21, 153
Stevenson, Thomas, 43
Stoeckl, Edward de, 19
Stokes, G. G., 103
Stokowksi, Leopold, 11–13
Stolberg, Louise von, 119–21, 156, 177
Stormont, Lord, 168
Stosch, Baron von, 119
Strawberry Hill Gothic, 128, 176
stroboscopy, 33
Stuart, Gilbert, 57
Suez Canal, 8, 218
sugar industry, 118, 252
Sullivan, Arthur, 41
sunglasses, 104
surgical gloves, 120–22
surgical techniques, 20–22, 24, 40–42, 153
Suvarov, Aleksandr, 38
Swift, Jonathan, 116
Swinton, Alan, 114
syphilis, 144

Taine, Hippolyte, 20, 31–33
Talbot, Fox, 70
Tales of Hoffmann (Hoffman), 81
talkies, 173
Talleyrand-Périgord, Charles-Maurice, 121, 131
Tambroni, Clotilde, 237–39
Taylor, G. I., 53
Taylor, John, 7–9
TDK, 213
tea, 155–56, 157, 229
Teach, Edward "Blackbeard," 148
telegraph cables, 19, 21, 70, 133, 138–40, 223, 240
telegraph signals, 91, 140, 172, 221
telephone, 242
telescopes, 63, 80, 82, 145, 245, 247
television, 11, 114
Tennyson, Alfred, Lord, 10, 111, 160, 171, 211–13, 241
Thackeray, William, 40, 160, 241
Thayer, Sylvanus, 46
theological chronology, 77
thermometers, 242
thin-film properties, 132, 134
Thirlwall, Connop, 8
Thomas, Lowell, 73
Thompson, Benjamin, 98
Thompson, Elizabeth, 110–12
Thomson, George, 59
Thomson, J. J., 173
Thornycroft, John, 193, 194
Thun, Countess, 207
Thurber, Jeannette, 91–93
Thurn und Taxis, Prince, 209
Ticknor, George, 61
Tilton, Theodore, 190
Tinsley, William, 41
Titanic, 53
Titus-Bode formula, 130
Todd, Charles, 221–23
Tompion, Thomas, 247
Tonson, Jacob, 199
torpedo boats, 193
Tournachon, Gaspard-Félix (Nadar), 243
Townshend, Charles, 157, 246–48
Tracy, Benjamin, 190
Trafalgar, Battle of (1805), 26–27, 41
transatlantic cable, 19, 70, 133, 240
treadmills, prisoners on, 156–58
trench digger, automated, 133
trench warfare, 195–96
triangulation, 20–22
Trinity College, 77
Tripoli, U.S. attack on, 5–6, 7
Trollope, Anthony, 41, 48, 188, 241
Troubridge, Thomas, 107
tuberculosis, 42, 142, 144
Tuckerman, Joseph, 48
turbines, 33, 53, 54
Turgenev, Ivan, 31, 40
Turgot, Anne-Robert-Jacques, 157
Tutankhamen, tomb of, 50
Tyndall, John, 163

Union College, 141
Unitarianism, 48
universal knowledge, 77–79
Uranus, 130
urban poverty, 48
Ure, Andrew, 191
uric acid, 43
Ussher, James, 77
utopian communities, 131, 186, 188, 191, 238

vaccines, 55–56, 57
vacuum cleaner, 192
vacuum flask, 132
Vail, Alfred, 138
Vaillant, Auguste, 20
Vanderbilt, Cornelius, 190
Vanderlyn John, 170–72
Van Rensslaer, Stephen, 251
van't Hoff, Jacobus, 33
Varley, Cromwell, 240
Varnhagen von Ense, Rahel Levin, 6
Vatican, 110
Vauban, Sebastien le Presle de, 196
velocity-distance relation, 64
vending machines, 202, 204
ventilators, 166
Verdi, Giuseppe, 243, 249
Verne, Jules, 243, 244
Viardot, Pauline, 31, 40
Victoria, Queen of Great Britain, 10, 18, 200, 209, 211, 253
Victoria amazonica, 200
viscose, 152
vitamins, 44
Voltaire (François-Marie Arouet), 36, 79–81, 89, 130, 177, 199, 249

Wagner, Richard, 230
Wallace, Alfred Russel, 173
Wallace, Richard, 181
Walpole, Horace, 89, 128, 176, 177
walrus tusks, 75
War of 1812, 7, 71, 109, 227
Washington, D.C., architecture of, 69–71, 91, 98
Washington, George, 78, 135, 136, 137, 170, 187, 206
watchmaking, 247
Water Babies (Kingsley), 18
Waterhouse, Benjamin, 57
watermarks, 227
water resistance, 250–52
Watson, James, 224
Watt, James, 118, 127, 168–70, 217, 236, 250
Watteau, Antoine, 86–88
Weatherly, Fred, 41
Weaver, John, 197
Webster, Daniel, 251–53
Wedgwood, Josiah, 196, 198, 235–36, 237
Wells, H. G., 23, 153
Wentworth, Paul, 168
Werner, Abraham, 129
Wesley, Charles, 226
Wesley, John, 226
West, Benjamin, 57, 250
Western Union, 138, 240
Westphalia, 67–69
West Point, 38, 46
Wharton, Thomas, 87
Wheatstone, Charles, 221
Whewell, William, 109, 111
whisky distillation, 217
Whistler, James McNeill, 21, 153
Whitbread, Sam, 109
White, Andrew, 131–33
Whitefield, George, 226, 227
white-light radiation, 34
white supremacy, 70
Whittle, Frank, 54
Wieck, Friedrich, 31
Wieland, Christoph Martin, 149–51
Wiener, Norbert, 73, 74
Wilberforce, William, 37, 186
Wilde, Oscar, 243
"Wild Peter," 115–16, 117
Wilkinson, John, 168
Wilkinson, William, 168–70
William Augustus, Duke of Cumberland, 248
William III, King of England, 86, 87, 148
Willis, Robert, 101–3
Winckelmann, Johann, 117–19
windmills, 230, 232
Wittgenstein, Ludwig, 231
Woffington, Peg, 166
Wolfe, James, 250
Wolff, Caspar, 249
Wollaston, William Hyde, 189–91
Wollstonecraft, Mary, 8, 219
women's rights, 171, 190, 210, 211, 219
Woodhull, Victoria Clafin, 190
Woodland, Norman, 174
Woolnoth, William, 68
Wordsworth, William, 8, 10, 219, 239, 249
World War I, 71, 193
World War II, 23, 73
worms, regenerative capability of, 79–81, 147, 149
Worth, Charles, 90
Wright, Frances, 131, 186–88
Wright, Thomas, 101

xenon, 33, 34
X-rays, 70–72, 212, 213, 223, 224

Yale College, 37
Yearsley, Anne, 185–86, 187
yerba maté, 229

Zola, Émile, 20, 218